PLASTIC FILMS

Technology and Packaging Applications

PLASTIC FILMS

Technology and Packaging Applications

Kenton R. Osborn
Formerly Technology Manager
Packaging Products Division
Polymer Products Department
E.I. du Pont de Nemours & Company, Inc.

Wilmer A. Jenkins
Formerly Director
Packaging Products Division
Polymer Products Department
E.I. du Pont de Nemours & Company, Inc.

IN COOPERATION WITH

Institute of
Packaging
Professionals

CRC Press
Taylor & Francis Group
Boca Raton London New York

CRC Press is an imprint of the
Taylor & Francis Group, an **informa** business

Published 1992 by CRC Press
Taylor & Francis Group
6000 Broken Sound Parkway NW, Suite 300
Boca Raton, FL 33487-2742

© 1992 by Taylor & Francis Group, LLC
CRC Press is an imprint of Taylor & Francis Group, an Informa business

First issued in paperback 2019

No claim to original U.S. Government works

ISBN 13: 978-0-367-45020-5 (pbk)
ISBN 13: 978-0-87762-843-9 (hbk)

Visit the Taylor & Francis Web site at
http://www.taylorandfrancis.com

and the CRC Press Web site at
http://www.crcpress.com

Library of Congress Cataloging-in-Publication Data

Main entry under title:
Plastic Films: Technology and Packaging Applications

Library of Congress Card Number 91-66456

CONTENTS

PREFACE

The commercialization of cellophane in the 1920s revolutionized the flexible packaging of consumer goods. For the first time, the buyer could see the contents of the package through a film that protected the packaged items from dirt, moisture, and atmospheric gases. Countless items previously packaged in heavy metal or fragile glass containers began to appear in this safe, convenient, light weight film. As a result, the flexible packaging industry grew from a small paper-based operation into the $5 billion giant it is today.

Cellophane rapidly grew to a multi-hundred million dollar business in the U.S. until the commercialization of high polymers in the early 1950s provided packagers with equally transparent films having better properties at lower cost. By the 1980s, plastic films had replaced 90% of the cellophane in flexible packages while plastics' versatility led to further growth in the use of lightweight, convenient, safe flexible packages as alternatives to traditional materials and packaging methods.

This book presents a comprehensive description of plastic films and their use in packaging. The reader will find a heavy emphasis not only on "what" and "how" but also on "why". The first three chapters are devoted to the technology of plastics and plastic films. Chapter 1 introduces the reader to polymer technology: what polymers are, how they are made, and how their properties are related to the structures and interactions of these long chain molecules. The second chapter describes the processes used to convert high polymers to thin, clear, strong plastic films. The third chapter completes the presentation of technology by describing how films are further processed to provide the properties needed for packaging applications.

The discussion then turns to those applications, beginning in Chapter 4 with a comprehensive description of the requirements of a package and the properties that any film must possess in order to meet these requirements. Chapter 5 deals with the complex world of packaging machinery: the equipment that has been developed to combine a plastic film with the product being packaged in it and the properties a film must have to be success-fully used on that equipment. With this background information in place, Chapter 6 then describes the plastic films that are used to package the mil-lions of items that are now packaged in this way, relating the requirements of the contents to the properties of the films that are used.

Finally, in Chapter 7 the authors look into the future and describe what they believe to be the likely evolution of plastic film packaging, with a dis-cussion of the major factors, both positive and negative, that will shape that future.

The Appendix contains a Glossary of Terms, Abbreviations and Test Methods; a Table of Film Properties; information on the current prices of plastic resins and films; and a description of how this complex industry is organized.

A thorough understanding of the material in this volume, particularly the technology presented in the first three chapters, requires that the reader be generally familiar with the elements of chemistry and physics presented in introductory college courses in those subjects. Additional technical terms have been defined as they appear.

The information in this book should be of value to many audiences:

- employees of companies that package their products for distribution and sale
- specialists and managers in companies who manufacture plastic films and resins for packaging converters and end users
- flexible packaging converters who want to learn more about the plastic films they use
- college students pursuing a career in packaging
- those working in the field who feel they need a broader comprehension of its many aspects as well as a reference source to help them keep up-to-date with this rapidly evolving technology

Although this book is designed to be a comprehensive treatment of the subject, three criteria have led to the deliberate exclusion of material that could have been included.

First, the focus is primarily on technology and applications current in North America. Modern film technology is much the same throughout the industrialized world, but cultural tastes and product consumption habits, particularly in food, vary from region to region, leading to significant re-gional differences in packaging methods.

Second, the material on applications attempts to concentrate on package types that are durable rather than transitory. As any supermarket shopper knows, packages come and go, changing as manufacturers attempt to gain market share by differentiating their offerings from those of their competi-tors. Any attempt to exhaustively describe *all* current packages would be tedious and soon outdated.

Third, rapid literary obsolescence has been avoided by dealing with process economics in a semi-quantitative, rather than a fully quantitative, manner. Thus, while absolute cost numbers are quoted in some cases, more often the cost of various processes are compared using phrases like "about twice" or "roughly three times" or "at least an order of magnitude greater than" rather than quoting actual numbers which unpredictable future events would soon leave outdated. The reader will find tables in the Appendix that show both plastic resin and plastic film prices. As time passes, these numbers will no longer be accurate in the absolute sense, but should have more enduring value as a portrayal of the relative price positions of these materials. The reader who needs precise cost information must always turn to suppliers for it rather than trust this or any other book, no matter how recent.

Many sources in addition to the authors' personal knowledge and experience have been used to compile the information presented here: packaging experts in many companies, trade literature, standard reference works, and other books that contain relevant, up-to-date material. The sources used will be found as footnotes on the pages where referenced and in the Bibliography section at the end of each chapter.

The authors are deeply grateful to the many individuals who made their time and knowledge available to help put this story together. The truth that appears here is often due to them, but never should they be held responsible for the inevitable errors and omissions — for these the authors are solely responsible.

BIOGRAPHIES

Wilmer A. Jenkins, Ph.D., received his Bachelor's Degree from Swarthmore College in 1949 and his Ph.D. in Chemistry from the California Institute of Technology in 1952. He then joined the DuPont Company as a research chemist, and went on to hold a variety of technical, manufacturing and sales positions in the fields of pigments, explosives, polymer intermediates and plastics. For the ten years prior to his retirement in 1989 he was Director of the Packaging Products Division of the Polymer Products Department. This division manufactured films (cellophane, Mylar®, Clysar® shrink films) and resins (EVA, ionomers, acid copolymers, HDPE, LDPE) for the flexible packaging industry. He was active in the Flexible Packaging Association during that period, serving on their Board of Directors and Executive Committee. Since retirement, he has been active as a consultant and co-author of books on plastics in food packaging and plastic films.

Kenton R. Osborn, Ph.D., received his Bachelor's Degree from Drew University and his Ph.D. in Physical Chemistry from the Illinois Institute of Technology. He joined the DuPont Company in 1955 as a research chemist in the Film Department and went on to hold a series of management positions in research and manufacturing. He was directly involved in all stages of the development and commercialization of new film products and processes, as well as the continual upgrading of technology. He has had first-hand experience with most of the commercially important products, including blown and oriented polyethylene films, oriented polypropylene film, oriented polyester film, polyvinyl fluoride film, polyimide film, cellophane, cellulose acetate film, shrink films, coated and coextruded films based on many of the above and numerous experimental films. For the fifteen years prior to his retirement in 1989, he was Technology Manager, directing all of the technical activities for DuPont's packaging products, including research and technical support for manufacturing and sales.

®Mylar and Clysar are registered trademarks of E. I. du Pont de Nemours & Co., Inc.

CHAPTER 1

Polymers

INTRODUCTION

There is an almost bewildering array of plastic films for packaging that are commercially available today. Their evolution was driven by market need on the one hand and by advances in technology on the other. These technological advances included development of both the number of polymers that are the current components of plastic films and the variety of processes for converting these polymers into films and film structures. In this chapter, an understanding of these polymers and how they are made will be developed.

Polymers are very long molecules, and while a few of commercial interest are formed in nature, like cellulose, most are synthesized using a chemical process called polymerization. In this process small molecules, called monomers, are joined together, end to end, in a growing chain. Eventually, the length of the polymer chain reaches a point where commercially useful objects can be made from them. Films made from short chains are brittle, but above some critical length they become tough and ductile. This transition point is different for different classes of polymers, but for polyethylene commercially useful chains have a minimum of 1000 to 2000 monomer units. Thus, the most important characteristic of polymer molecules is their length since that determines whether they have any usefulness at all.

The next most important characteristic is the chemical nature of the units making up the polymer molecule. Degrees of toughness, stiffness, transparency, barrier to gases, etc. exhibited by films of chemically different polymers vary widely. Additional variation is also caused by the specific conditions under which the polymerization takes place. Also, combinations of monomer units can be caused to polymerize together forming a copolymer. Finally, different polymer molecules can be mixed together or blended leading to further proliferation of the levels and combinations of properties that are achievable in plastic films.

1

In the next sections, an understanding of polymer molecules and their structure will be developed and used to explain their fundamental properties. The fundamental properties of polymers are those that are important regardless of how they are used; the properties of films will be described in Chapter 2 where film processes and products are discussed.

MOLECULAR STRUCTURE

The simplest member of the family is polyethylene. It is made up of repeating units with the chemical structure CH_2, as in Figure 1.1.

Much of the discussion that follows on polymer structure and its relation to properties will be based on polyethylene since it serves as a good model for polymers in general. Later in the chapter, the effects of the chemical structures of other more complex polymers will be covered. Polyethylene is made from ethylene gas. The process will be described in detail later but for the purposes of this discussion can be simplified as follows. When the ethylene molecule is attacked by a free radical (designated as $R \cdot$), the free radical bonds to the ethylene and transfers the free radical site to the ethylene end of the new molecule, as in Figure 1.2. The new molecule is still a free radical, so it can attach itself to another ethylene molecule, again transferring the free radical site to the growing end of the molecule. This process of repeated additions is the dominant one because of the high concentration of ethylene molecules and the small concentration of free radicals. However, two free radical ends occasionally meet and join together completely eliminating the free radical nature of the two molecules. When that happens, the polymer molecule ceases to grow.

Other chemical reactions also terminate the growth of the polymer molecule. These involve reactions of the free radical entities with chemically active sites on the polymer molecule or with other chemically active molecules that are present as impurities or are deliberately added to control the length of the polymer molecule. Thus, for one reason or another, the polymer chain does not grow indefinitely. Finished polymer chains vary in length since the events that compete with the polymerization process are controlled by probabilities that depend on relative rates of the various possible chemical reactions and on the relative concentrations of

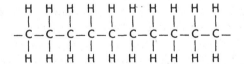

Figure 1.1 The Structure of Polyethylene.

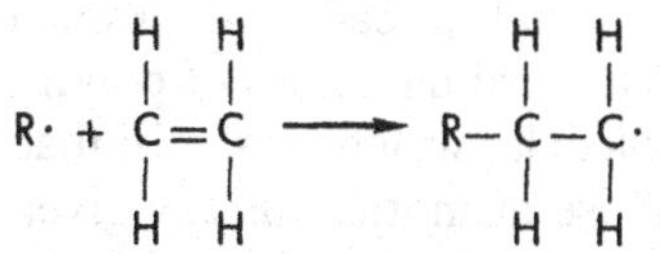

$$R \cdot + \underset{\underset{H}{|}}{\overset{\overset{H}{|}}{C}} = \underset{\underset{H}{|}}{\overset{\overset{H}{|}}{C}} \longrightarrow R - \underset{\underset{H}{|}}{\overset{\overset{H}{|}}{C}} - \underset{\underset{H}{|}}{\overset{\overset{H}{|}}{C}} \cdot$$

Figure 1.2 The Reaction of Ethylene with a Free Radical.

the reacting chemical species. As a result, there is always a distribution of polymer lengths.

Molecular Weight and Molecular Weight Distribution

The length of a linear polymer chain is usually described by its molecular weight, which is simply the molecular weight of the monomer multiplied by the number of monomers in the polymer chain. For polyethylene, the molecular weight of the monomer (CH_2–CH_2) is 28 and a common number of these units is about 5,000. Thus a typical molecular weight for polyethylene is 140,000. Commercial polyethylenes range in molecular weight from about 50,000 to over 1,000,000.

The distribution of molecular weights for polymers is described by curves similar to the idealized Gaussian curve shown in Figure 1.3. This curve is constructed by plotting the molecular weight along the x axis and the number of molecules having that molecular weight on the y axis. How the curve is determined experimentally will be described later. If the curve had always exactly the same symmetrical shape as defined in Figure 1.3, then a single number, defined by A, the intersection of the axis of symmetry with the curve (as shown below), would suffice to describe the distribu-

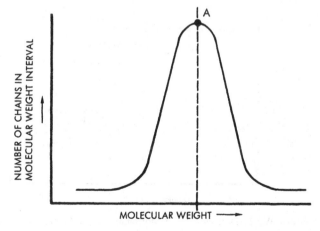

Figure 1.3 Idealized Curve of the Molecular Weight Distribution for Polymers.

tion. This single number could be called the average molecular weight. Everyone would know that equal quantities of polymers would be present with higher and lower molecular weights and, for that matter, could define from the curve exactly those quantities for any given range of molecular weights.

In reality, however, there are two complications. The first is that the distributions are not symmetrical and the shapes of the distribution curves depend on the process used to make the polymer. A curve for a typical polymer is shown in Figure 1.4. Note here the existence of high molecular weight "tail": that is, a much higher fraction of high molecular weight polymer molecules than would be the case for a symmetrical distribution curve. This, and a low molecular weight tail, are both common occurrences in the polymerization process. The presence or absence of such tails can have a subtle but significant effect on the physical properties of polymers and hence, it is important to be able to characterize the distribution curve.

The second complication is that different methods of measuring an average molecular weight can lead to different numbers as shown in Figure 1.4. For example, one number is a number average molecular weight (M_n). This is derived from the total weight of polymer molecules divided by the total number. Another kind of number is a weight average molecular weight (M_w) which is determined by multiplying the weight of each chain of a given number of repeat units by the number of such chains and dividing by the total weight of chains. When both of these are known, they can be used to characterize the shape of the distribution curve. The ratio, (M_w/M_n) is used to judge whether the distribution is narrow or broad. For example, a ratio of 6 is considered a narrow distribution for polyethylene while a ratio of 12 is considered a broad distribution.

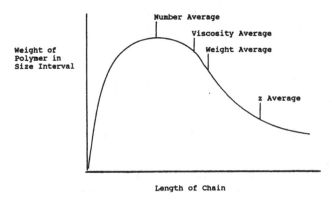

Figure 1.4 Molecular Weight Distribution and Molecular Weight Averages in a Typical Polymer.

Measurement of Molecular Weight and Molecular Weight Distribution

The behavior of a polymer molecule in the solid or melted state is influenced heavily by interactions with neighboring molecules. Therefore, measurements that reflect the characteristics of an isolated molecule (such as its molecular weight) are made in very dilute solutions. One of the traditional measurements of molecular weight is based on the osmotic pressure of dilute solutions. This pressure is the differential pressure sustained across a membrane that separates the pure solvent from a dilute solution of a polymer and that permits the passage of solvent only. The osmotic pressure is a function only of the number of molecules in solution. Therefore, knowing the total weight of polymer, a number average molecular weight can be calculated.

Another determination derives from light scattering measurements using a wave length of light that approximates the dimensions of the polymer molecule. This leads to a weight average molecular weight and represents the statistically expected size of the average molecule. As such it is very sensitive to the presence of larger molecules. Measurement of the rate of sedimentation in a dilute solution subjected to high centrifugal forces will also yield an average molecular weight. This is the "z average" molecular weight shown in Figure 1.4.

All of the above techniques are difficult and tedious and not suitable for the routine measurements needed to control a commercial process. Therefore, simpler measurements that correlate empirically with molecular weight are employed. One of these is the measurement of viscosity in dilute solution. A correlation can be established between viscosity and molecular weight from the viscosities of samples of known molecular weight and with a narrow molecular weight distribution. This "viscosity average molecular weight" falls in between the number and weight average molecular weights as shown in Figure 1.4.

As mentioned earlier, some indication of the breadth of the molecular weight distribution curve can be made if both the number average and weight average molecular weights are known. A more complete description can be established by fractionation of the sample using a technique that depends on the size or weight of a molecule. A quick and convenient technique for this determination is gel permeation chromatography or GPC. Here a solution of a polymer is passed through a column consisting of beads of porous glass, or cross-linked polymer containing pores of a controlled size. Smaller molecules from the solution penetrate and are retained for a time in the pores while larger molecules pass directly through the column. Thus at the exit of the column the highest molecular weight fraction appears first followed by fractions of decreasing molecular weight. The concentration of polymer in the exiting solution is measured by changes in the index of refraction or by the absorption of UV light. These measurements can lead to a relative molecular weight distribution or, by calibration with fractions of known molecular weight, to an absolute measure of molecular weight and molecular weight distribution.

The Effect of Molecular Weight and Molecular Weight Distribution on Physical Properties

As the molecular weight of polyethylene increases, melt viscosity, tensile strength, impact strength, abrasion resistance, and shrinkage at elevated temperature all increase. However, processability is more difficult and there is a deterioration in optical properties with increasing haze and decreasing gloss. The characteristic shape of the curve relating molecular weight to many of these properties is shown in Figure 1.5 for polyethylene. Here, melt flow index is the measure of molecular weight, increasing as molecular weight decreases, and the strength property is the energy required to break a strip of film stretched at room temperature over a range of rates of strain. Note that energy to break at low rates of strain $(3.3 \times 10^{-4}$ and $8.3 \times 10^{-3})$ increases and then levels off at an intermediate molecular weight. Similar effects would be seen for the elongation to break, tensile strength, and modulus. In contrast, at the higher rate of strain the energy to break tends to keep increasing up to very high molecular weights. This same trend is typical of properties like puncture resistance and impact strength where similarly high rates of strain are involved.

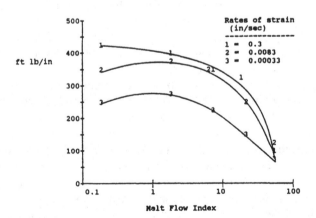

Figure 1.5 The Relationship between Melt Index and Energy to Break for Polyethylene.[1]

[1]Renfrew, A., ed. 1957. *Polythene – The Technology and Uses of Ethylene Polymers*. New York, NY: Interscience Publishers, Inc., p. 179.

All of these properties are clearly due to the length of the polymer molecules since crystallinity, which will be discussed later, is essentially constant in these samples. The advantageous effect of length is to increase the number and degree of entanglements and the intermolecular attraction within the network of polymer molecules that form in the solid state. Polymers of commercial interest are generally made in the intermediate molecular weight range where physical properties have leveled off for the most part. Special high molecular weight grades can be made but are usually more costly to produce and more difficult to convert into films.

Compared to chain length, molecular weight distribution is a secondary effect. For polyethylene, as molecular weight distribution broadens, melt flow increases at a given applied force but melt strength, tensile strength and impact strength decrease modestly. There is also a tendency for surface gloss to decrease.

Branching in Polymer Molecules

So far, for the sake of simplicity, it has been assumed that polyethylene molecules made by a free radical polymerization process are completely linear as depicted in Figure 1.1. In fact, this is not the case. It was mentioned earlier that a way in which the growing polymer molecule is terminated is by the free radical entity on the end attacking a chemically active site somewhere along a polymer chain. When this occurs the free radical activity is transferred to that site allowing a new polymer chain to grow out from the original molecule at that site. By this mechanism, branched molecules are formed as shown in Figure 1.6. The hydrogen atoms have been left out so that just the carbon-carbon backbone of the molecule is shown. Also, E is used to indicate a terminated chain end. In this example, one branch is quite short while the second is very long, perhaps approaching the length of the original molecule. Since the branches are initiated and grow at the same time that the polymer molecule is growing, they are subject to the same competitive processes as the main chain

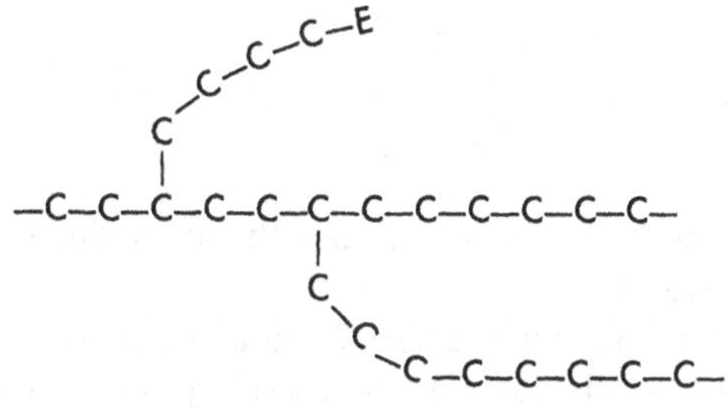

Figure 1.6 Branches in Polyethylene.

leading to a distribution of lengths of branches. Both this distribution and the average chain length are determined by the conditions of polymerization. If there is a high concentration of free radicals in the system more branches will be formed and both they and the main chain will be shorter. As the free radical concentration is decreased, the length of the main chain and the branches increase and the frequency of branching decreases. For polyethylene made by the free radical process, the number of branches per polymer molecule is in the range of 30 to 40 with most of these being very short branches of 2 to 4 carbon atoms each and with just a few long chain branches. For linear polyethylene made by a different process the number of branches is only 4 to 5 per chain.

While direct measurement of the degree of branching is difficult, infrared spectrometry can be used for many polymers to measure the number of end groups. However, concentrations of the end groups are often very small, making quantitative measurement subject to significant error. On the other hand, branching can be inferred by its effect on crystallinity and on the rheology or flow characteristics of the polymer melt. These two topics will be discussed next.

THE RHEOLOGY OF POLYMER MELTS

Understanding the flow of a liquid polymer, whether to gain structural insights or to aid in the design and control of equipment to transform polymers into plastic films, depends critically on defining how the melt responds to the applied forces. This response is characterized by the strain or, more importantly, the rate of strain resulting from a continuously applied stress. The simplest case is where the melt is deformed and caused to flow between two parallel plates as shown in Figure 1.7. The upper plate is moving at a velocity V along the x axis while the lower plate is stationary. The velocity of the fluid is variable depending on where it is measured along the y axis. The rate of strain is defined as the rate of change of this velocity and is equal to the slope, dv/dy. Since the deformation of the fluid is a shearing one, dv/dy is called the shear rate, designated hereafter as $\dot{\nu}$, and the force applied to upper plate is the shear stress, τ. Newton's law for steady, laminar flow in this situation is:

$$\tau = \eta \cdot \dot{\nu} \tag{1}$$

where η is the viscosity, a constant. Fluids that obey this simple relationship are called Newtonian.

While there are instruments that simulate the ideal situation in Figure 1.7, the common way to characterize the rheology is to study the flow of a fluid through a capillary with a well defined geometry and under carefully

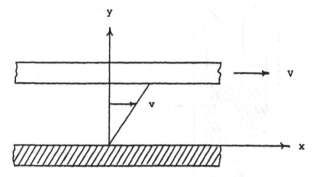

Figure 1.7 Melt Flow between Parallel Plates.[2]

controlled temperature using an arrangement such as that shown in Figure 1.8. Here, the piston is driven with a constant force and the rate of flow is measured. From the dimensions of the capillary, the rate of shear is calculated from the rate of flow. This rate of shear is the velocity gradient from the center, where the velocity is a maximum, to the wall of the capillary where it is a minimum. At high flow rates the profile is very elongated and the shear rate is high. At low flow rates the profile is nearly flat and the shear rate is correspondingly low. The pressure being applied to cause flow can be converted into a shear force using the geometry of the capillary.

Flow data obtained from such an instrument are shown in Figure 1.9 for polyethylene measured at 126°C where viscosity is given as a function of the shear rate. The apparent viscosity, though constant at very low shear rates, decreases sharply at higher shear rates. Such behavior is called non-Newtonian and is caused by the change in the shape of the polymer molecules as they are subjected to higher and higher shear rates. In the undisturbed melted state, the polymer molecule is a random coil highly entangled with other coiled molecules like a mass of cooked spaghetti. This accounts for the very high viscosity of polymer melts. For any flow to occur, the movements of each molecule must be coordinated with those of many neighbors. As this entangled mass is subjected to a shearing force the molecules begin to pull away from the entanglements and to be aligned more and more with their neighbors. Thus the resistance to flow gradually decreases. At very high shear rates, the molecules are almost totally aligned in the direction of flow and interaction with their neighbors approaches a very low value. This high degree of order is an unstable state

[2]McKelvey, J. M. 1962. *Polymer Processing*. New York, NY: John Wiley and Sons, Inc., p. 12.

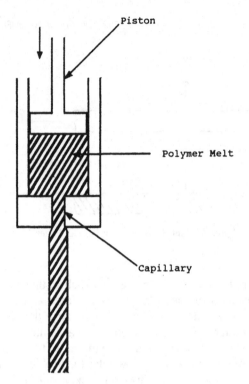

Figure 1.8 Capillary Rheometer.

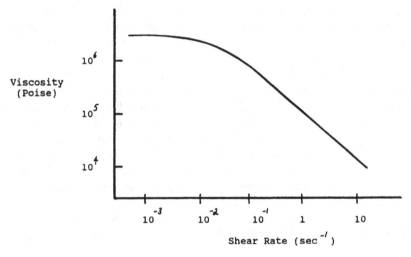

Figure 1.9 Melt Rheology of Polyethylene (measured at 126°C).[3]

[3]Ibid. p. 34.

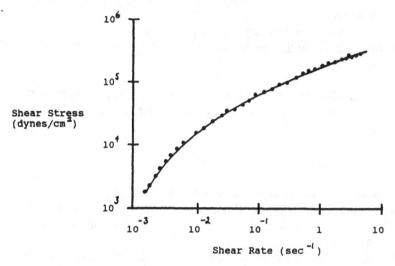

Figure 1.10 Logarithmic Flow Curve for Polyethylene at 126°C.[3]

and the molecules will return to their normal coiled, entangled condition as the melt is allowed to recover from the experience of flowing under high shear force. In many cases, this recovery of the normal state causes a swelling of the melt as it emerges from the capillary.

The structure of a polymer molecule strongly affects its rheology. The melt viscosity at low shear rates increases with the weight average molecular weight. Long chain branching of the polyethylene molecule and a broader molecular weight distribution cause the viscosity to be more sensitive to shear rate.

Attempts to derive equations of flow for non-Newtonian fluids from theory have led to some very complex results that are difficult to use. Fortunately, a simple empirical equation describes the relationships quite well for a wide range of polymers over a fairly broad range of shear rates. This equation results from the observation that when the log of shear rate is plotted against the log of shear stress a straight line is found at least for significant portions of the data as shown for LDPE measured at 126°C in Figure 1.10. Data like this can be represented by the following relationship:

$$\eta = \eta' \left(\frac{\dot{\nu}}{\dot{\nu}'} \right)^{n-1} \qquad (2)$$

where

η = viscosity

$\dot{\nu}$ = shear rate
$\dot{\nu}'$ = shear rate in an arbitrary state, usually 1 sec^{-1}
η' = the viscosity at that shear rate
  n = flow index

For the region between 0.1 and 1 sec^{-1} in the figure above, n is 0.59. For $\dot{\nu}'$ equal to 1 sec^{-1}, η' is 1.5×10^5 dynes/cm^2 and the viscosity equation becomes:

$$\eta = 1.5 \times 10^5 \, (\dot{\nu})^{-0.41} \tag{3}$$

Temperature is also an important variable. For polyethylene, a temperature rise in the melt of about 60°C decreases the viscosity at low shear rate by a factor of 10. However, this temperature effect is greatly reduced at high shear rate.

Although the flow of a polymer melt through a capillary is often used to characterize rheology, there are limitations in this approach. For one thing, the melt does not instantaneously convert from the shape of the cylinder (see Figure 1.8) to that of the capillary. There is a transition zone above the capillary which in effect extends the length of the capillary so that "entrance corrections" must be applied to the calculations. There are other instruments designed to overcome this problem by shearing the melt between two discs or between a plate and a cone approaching more theoretically rigid conditions. These instruments also have the advantage of being able to make measurements at very low shear rates which are necessary to define more quantitatively the elastic properties of the melt.

By elastic properties is meant the tendency of the melt to recover or to return to its original shape after it has been deformed. An example is the swelling of the melt as it exits a capillary as described earlier. In a sense, the melt remembers its original form or shape. The more entangled a network of polymer molecules, the more elastic will be its behavior. An extreme case would be a highly cross-linked network (physical links between the molecules). Melts that exhibit an elastic component of flow are said to be viscoelastic. One can imagine that if a highly entangled network is deformed only slightly that this deformation is totally recovered once the force is removed. This is readily shown in a cone and plate instrument where the cone resting in the melt is rotated slightly. Upon release of the force on the cone it returns to its original position. As the force is increased, some viscous flow takes place as well so the total deformation is not fully recovered.

If there is a high degree of elasticity in a melt, it may take extended "shear working" to achieve the degree of disentanglement of the network that is characteristic of the shear rate being applied. In extreme cases this can be seen as a decrease in the melt viscosity each time it is measured by quickly recovering and reprocessing the sample when it emerges from a capillary. This is an important factor in the design of the conditions for extrusion of polymers, as will be described in Chapter 2.

The properties of polymer melts stem from the principle that the stable state for a polymer molecule in a molten condition is a random coil highly entangled with its neighbors. For a solid polymer, however, the stable state is a more ordered one. Many polymers form crystalline domains similar in many respects to crystals formed from simple compounds like salt. The nature of these crystalline domains is highly sensitive to the structure of the polymer molecule.

CRYSTALLINITY IN POLYMERS

Crystalline and Amorphous Domains

As a polymer melt is allowed to slowly cool, the molecules tend to align themselves with their neighbors. Attractive forces between the molecules are re-enforced as the alignment becomes increasingly symmetrical. There is recent evidence that molecules also fold on themselves in parallel arrangement and that, in fact, this folding may be the dominant mechanism, with order increasing as neighboring molecules align themselves and fold on themselves. Whether by intermolecular alignment or intramolecular folding, as ordering approaches perfection and the melt begins to approach solidification, crystalline domains form. In these domains, the repeating units of the polymer chains occupy specific sites that form a symmetrical geometric solid. An example would be a tetrahedron with one unit in each corner. These crystalline domains grow to be quite large, often reaching the dimensions of a wave length of light, so that the resulting scattering of light causes haziness in films. Within a crystalline domain, units occupying neighboring sites can belong to different polymer chains. Thus, any given polymer molecule may wander in and out of several different crystalline domains. A large volume, containing many of these domains, consists of highly ordered and condensed crystalline areas with more open, disordered, or amorphous areas in between. These more disordered regions inevitably occur because as a polymer chain is being incorporated into a crystalline domain a point is reached where the next part of the molecule will not fit into the crystalline symmetry either because the molecule is so highly entangled with others or because some disruption such as a branch occurs. Thus, high molecular weight polymers are never completely crystalline and generally fall into the range of 20 to 90% crystallinity.

Response to Temperature

Crystalline domains increase in perfection by annealing. This implies that even in the solid state, polymer molecules or at least segments thereof have sufficient mobility at a certain temperature to be able to assimilate themselves better into a crystalline matrix. The initiation of this mobility occurs at the glass transition temperature (T_g) or the second order transition temperature. In a commonly used method, this temperature is determined by measuring the temperature difference between the polymer sample and an inert reference material as the temperature of the system is slowly raised at a constant rate. The polymer sample and the reference will heat up at the same rate, up to the point where segmental motion introduces a new mode of energy absorption. At this temperature an inflexion occurs in the plot of the temperature difference versus the temperature of the reference material, as shown schematically in Figure 1.11. This technique is called differential thermal analysis (DTA). In another technique, called differential scanning calorimetry, the energy required to maintain the temperature of the polymer sample and the reference material at an equal level is measured.

The segmental motion that begins at T_g occurs largely in the amorphous regions since in the crystalline regions, movement is restricted. As the temperature of the sample is increased further, segmental motion continually increases, primarily in the amorphous regions. At a high enough temperature, however, the forces binding the segments together in the crystal matrix are overcome and the crystal melts. This is a very sharp transition as shown in Figure 1.11 (T_m). The crystalline melting point and

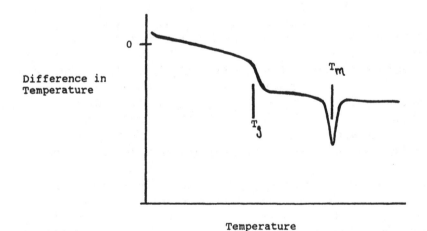

Figure 1.11 Schematic DTA Curve Showing Thermal Transitions.

the glass transition temperature are dependent largely on the chemical structure of the polymer chain.

Measurement of Crystallinity

As mentioned above, crystalline domains of polymers exhibit some of the same properties as crystals of simple molecules. One characteristic is that they diffract X-rays, with the diffracted beam forming a pattern on a photographic plate. Analysis of this pattern leads to precise information about the form and dimensions of the crystal as well as the relative amount of these versus the amorphous areas.

Another property of crystalline regions is that they are denser than amorphous regions. For example, the most crystalline form of polyethylene has a density of about 0.96 whereas the least crystalline form has a density of about 0.88. Thus, for polymers of similar chemical structure, degree of crystallinity can be inferred from a knowledge of the density. One way to measure density is to immerse a sample in liquids of different density seeking one in which it neither sinks or floats. For many routine measurements of polymers within a limited range of densities, a density gradient is established in a large, glass cylinder by carefully filling it with mixtures of miscible liquids with the right range of densities. After calibration, the density of any polymer sample can be exactly measured by noting the position it assumes in the cylinder.

The Effect of Molecular Structure

From the description of crystalline and amorphous domains in the beginning of this section, it follows that the most linear polyethylene molecule leads to the highest degree of crystallinity (about 90%) in solid form. Since it exhibits the highest density (0.96) it is called high density polyethylene (HDPE). HDPE is not produced by free radical polymerization and hence has a very low degree of branching. For polyethylene produced in the free radical process, with its much higher level of branching, both crystallinity (50–70%) and density (0.91–0.92) are lower. This variety of polyethylene is called low density polyethylene (LDPE).

Another way to introduce disruptions in the linearity of a polyethylene molecule is to include a second monomer during the polymerization which contains a bulky side chain. An example would be a copolymer based on ethylene and a small amount of butene as shown in Figure 1.12.

Here, both the length and number of the side chains or branches are precisely controlled. This polymer is made by the same process as is HDPE and is called linear low density polyethylene (LLDPE). Densities for LLDPE are typically in the range of 0.94 to 0.95. With bulkier side chains,

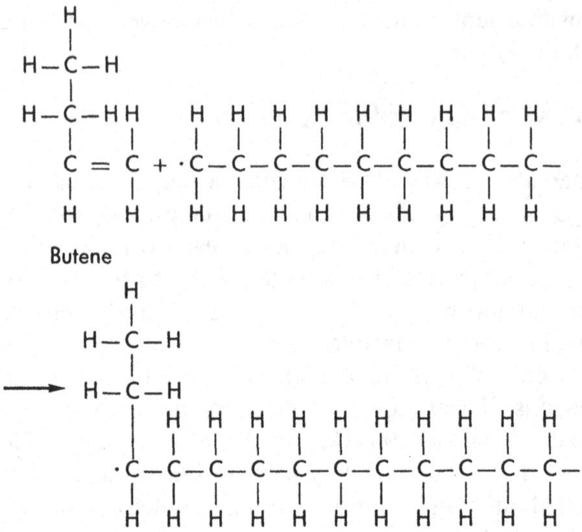

Figure 1.12 Copolymerization of Ethylene/Butene.

which are formed if a higher molecular weight hydrocarbon such as hexene is used as a comonomer, densities as low as 0.88 have been attained. These new varieties of polyethylene are called very low density (VLDPE) and ultra low density polyethylene (ULDPE). Varying the kind of branching produces subtle effects beyond controlling the amount of crystallinity. For example, in Table 1.1 it is seen that LLDPE can have a higher crystalline melting point than LDPE even though they have the same density.

By means of controlled branching, then, a wide range of degrees of crystallinity is possible. Other variations, to be described later, in the chemical structure of the polymer molecule also affect crystallinity. The ability to control crystallinity has significant practical implications since many of the important physical properties depend on degree of crystallinity.

Crystallinity and Physical Properties

The presence of crystalline domains decreases the mobility within a polymer network and makes it more resistant to an applied force. Thus,

Table 1.1. Comparison of LLDPE and LDPE.

Polymer	Density (grams/cm^3)	Crystalline MP (°C)
LLDPE	0.92	118
LDPE	0.92	95

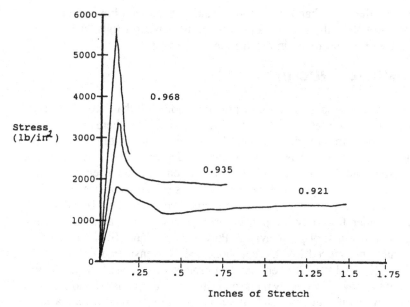

Stress, (lb/in²)

Inches of Stretch

0.968

0.935

0.921

Figure 1.13 Tensile Curves for Polyethylenes of Different Densities[4] (Extension rate 0.3 in/sec at 20°C).

with increasing degree of crystallinity a polymeric structure exhibits increased stiffness, increased tensile strength, and a lower elongation to reach the breaking point. For a plastic film, these properties are measured by pulling a strip to its breaking point under precisely controlled conditions: dimensions of the strip, rate of stretch and temperature. Typical data from such a measurement for a variety of polyethylenes is shown in Figure 1.13. Stiffness as measured by the tensile modulus is the slope of the initial part of these curves, whereas tensile strength and elongation are the intersections of the break point on the yield stress and extension axes respectively. Note how the modulus and tensile strength increase with increasing crystallinity and density, while elongation decreases.

An increase in rigidity and strength is desired in some applications; in others a more ductile, flexible structure is required. As noted above, with increasing crystallinity film haze increases because of the scattering of light by the crystallites. On the other hand, the barrier to the diffusion of gases increases with increased crystallinity. Thus, crystallinity must be optimized to meet the needs of each application and for the processing method to be used. More crystalline polymers tend to have higher melting points and must be processed at higher temperatures where melt strength is undesirably lower and where thermal degradation may become a prob-

[4]Renfrew, p. 173.

lem. On the other hand, the more crystalline, stiffer films are easier to move through a film process and to wind into uniform rolls. All of these effects will be discussed in detail in later chapters.

CHEMICAL STRUCTURE

So far, the variations within a polymer molecule that have been considered have been restricted to polyethylene molecules where only carbon-carbon and carbon-hydrogen bonds exist. Even so, within this limited range of polymers it has been possible to describe all of the important properties and their relationship to structure on both the levels of a single molecule and of a system of molecules. We shall now consider the effect of chemical entities other than carbon and hydrogen. This broader look will consider first only the pendent groups attached to a carbon-carbon backbone like that of polyethylene. Polymers that have different chemical structures introduced into the backbone will be considered next.

To understand how different pendent groups affect crystallinity, it is important to know how readily the group can be accommodated into a crystal lattice. A small group is much more readily accommodated than a large one. An example of a small pendent group still from the family of polyhydrocarbons is polypropylene (see Figure 1.14). The methyl groups along the chain fit well into a crystal matrix, leading to a relatively high level of crystallinity. Another example is a polymer made from vinyl alcohol (see Figure 1.15) which is highly crystalline.

A relatively small pendent group such as chlorine also yields a relatively high level of crystallinity as in polyvinyl chloride (PVC), which is shown in Figure 1.16. Another example of a small pendent group is cyanide and, indeed, polyacrylonitrile (PAN), shown in Figure 1.17, is highly ordered though not with traditional crystalline and amorphous regions. Increasing symmetry also helps as in the noteworthy example of polyvinylidene chloride (PVDC), depicted in Figure 1.18, which is highly crystalline. As the pendent groups get larger and bulkier, they reach a point where formation of a crystalline lattice is not possible; such polymers are amorphous. Two common examples are polystyrene (PS) with its bulky, pendent ben-

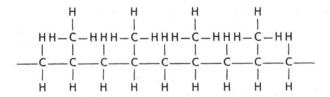

Figure 1.14 The Structure of Polypropylene.

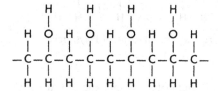

Figure 1.15 The Structure of Polyvinyl Alcohol.

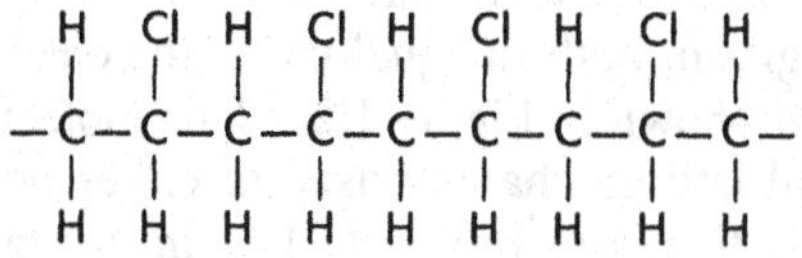

Figure 1.16 The Structure of Polyvinyl Chloride.

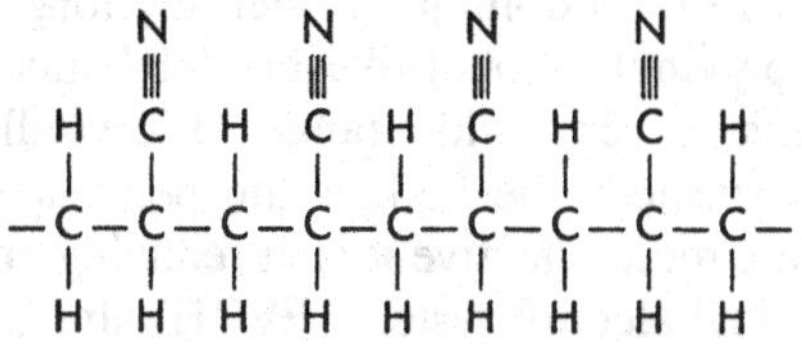

Figure 1.17 The Structure of Polyacrylonitrile.

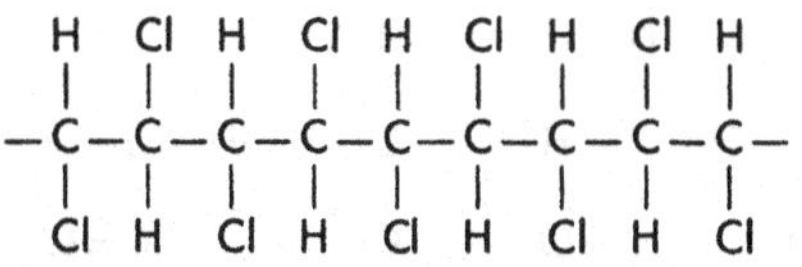

Figure 1.18 The Structure of Polyvinylidene Chloride.

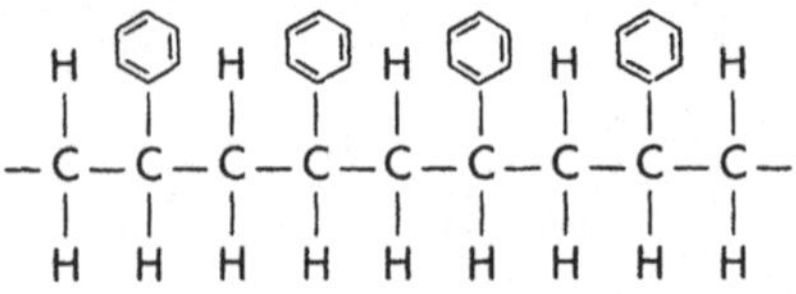

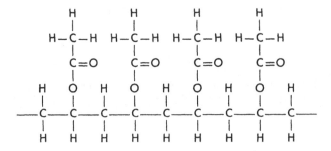

Figure 1.19 The Structure of Polystyrene.

zene ring (see Figure 1.19) and polyvinyl acetate (PVAc) with its large acetate group shown in Figure 1.20.

The examples so far have considered only homopolymers (polymers made from one monomer). Additional flexibility in the control over crystallinity can be attained using mixtures of monomers to make copolymers. For a copolymer with ethylene units randomly dispersed among units containing an OH group, both in equal parts, the chain would show an alternating structure as shown in Figure 1.21. This molecule would not fit exactly into the crystal lattices characteristic of either polyethylene or polyvinyl alcohol, and is therefore relatively low in crystallinity. If the polymer is mostly ethylene, the molecule would consist of long runs of polyethylene interspersed with sections of the copolymer, as above. The polyethylene sections would readily form crystalline domains so that the total system would be moderate in level of crystallinity. Similarly, if the unit containing the alcohol group dominates, then the long runs of these units would fit into a polyvinyl alcohol-like crystal lattice. Copolymer systems like these permit a very wide range of crystallinities to be achieved. As might be expected, the bulkier the pendent group on the modifying comonomer, the more effective it is in reducing crystallinity. In addition to the ethylene/vinyl alcohol system (EVOH) already mentioned, other copolymers important to plastic films are: ethylene/vinyl acetate (EVA), ethylene/acrylic acid (EAA), ethylene/methacrylic acid (EMAA), vinyl chloride/vinylidene chloride, and vinylidene chloride/acrylonitrile.

Figure 1.20 The Structure of Polyvinyl Acetate.

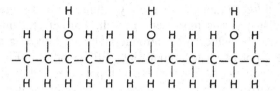

Figure 1.21 The Structure of an Alternating Ethylene/Vinyl Alcohol Copolymer.

So far, the effect of the size of a pendent group has been emphasized. The specific chemical interactions are also important since they can lead to increased intermolecular attraction. Groups like alcohol, cyanide, chloride, acid, and acetate are polar and tend to form hydrogen bonds, which represent the strongest link between polymer molecules short of an actual chemical bond. Increased attraction is an additional force, beyond crystallinity, which tightens the polymer network. This effect is shown in Table 1.2 where the T_g of a series of polymers is seen to increase with the increasing polarity of the pendent group. Polarity consequently increases the resistance to deformation by applied forces, increases the rigidity of the network and increases the barrier to the diffusion of gases. Also there are important chemical effects at the surface of a plastic film. All of these matters will be discussed in Chapter 2.

Another variable is introduced by varying the chemical structure of the polymer molecule: thermal decomposition. To begin with, as the melting point is raised by crystallinity and hydrogen bonding, the decomposition temperature of the polymer can be approached. In some cases, the polymer molecule will depolymerize. An example is polymethylmethacrylate, which can be decomposed completely back to monomer. In other cases, a reaction of the pendent groups can lead to the splitting out of a molecule. This is the case with chlorine-containing polymers where hydrochloric acid is evolved from the melt if the temperature is too high or if the time in the molten state is too long.

The carbon-carbon backbone, which is the only one considered so far, is relatively flexible. Bulky pendent groups interfere with that flexibility. For example, polystyrene, with its benzene ring on every other carbon

Table 1.2. The effect of polarity on T_g.

Polymer	T_g (°C)
LDPE	−25
PVAc	29
PVC	105

atom is fairly stiff even though it is noncrystalline. Also, as noted above, hydrogen bonding restricts movement as well. Further stiffening of the backbone can also be achieved by introducing larger, bulkier groups into it. An example is polyethylene terephthalate (PET), shown in Figure 1.22. This is not a copolymer of ethylene but is a polyester and is made by a different polymerization process called condensation which will be described later in this chapter. These polymers are quite stiff due both to their crystallinity and to the introduction of the benzene ring into the chain. The extreme of backbone stiffening is seen in liquid crystal polymers, in which the backbone is completely aromatic (all benzene rings). These molecules are like stiff rods even though they are amorphous. They also exhibit excellent resistance to the diffusion of gases. Unfortunately their high cost precludes them from interest as packaging films at this time.

POLYMERIZATION PROCESSES

Each addition of one monomer to another in the polymerization process is a chemical reaction. In some cases the nature of the monomer requires a different kind of reaction to be used, while in other cases a different reaction can be used to vary the structure of a given polymer. For these reactions to occur in an optimum way, each must be run under very different conditions (varying temperature, pressure, solution, dispersion, etc.).

Free Radical Process

A free radical is an unstable molecule that is generated by breaking a chemical bond. Compounds that generate free radicals at relatively low temperatures are called initiators, and an example is shown in Figure 1.23. The starting molecule on the left contains a peroxide linkage ($-O-O-$) which readily decomposes at a temperature that depends on the nature of the rest of the molecule (the R's). Sometimes a second chemical is used to cause decomposition, especially when polymerization at relatively low temperatures is desired.

Once free radicals have been formed, one possible subsequent reaction is for the two radicals to join and form a stable new compound. However,

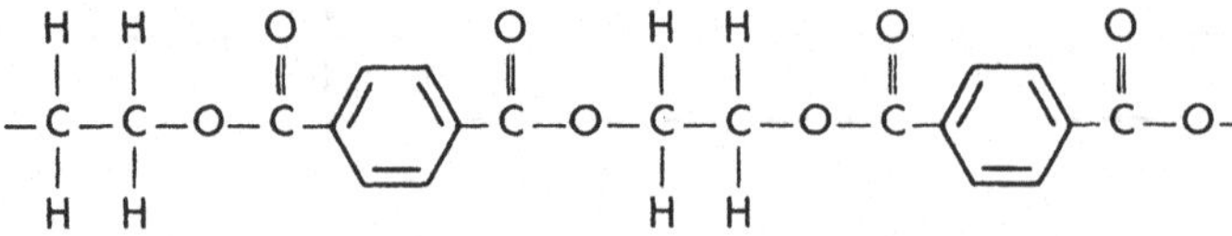

Figure 1.22 The Structure of Polyethylene Terephthalate.

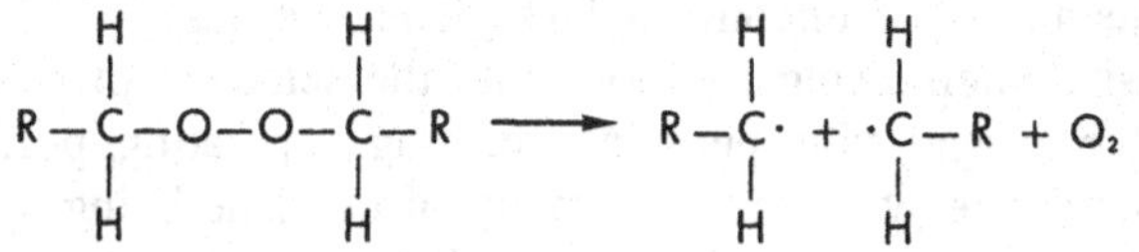

$$
\begin{array}{ccc}
\underset{\underset{H}{|}}{\overset{\overset{H}{|}}{R-C-O-O-C-R}} & \longrightarrow & \underset{\underset{H}{|}}{\overset{\overset{H}{|}}{R-C\cdot}} + \cdot\underset{\underset{H}{|}}{\overset{\overset{H}{|}}{C-R}} + O_2
\end{array}
$$

Figure 1.23 Free Radical Formation from a Peroxide Initiator.

in a typical polymerization reaction, the concentration of initiator is small relative to that of the monomer so that the dominant reaction is for the free radicals to attack a monomer molecule. Such an attack occurs readily on a double bond so that free radical initiation is often used to polymerize ethylene and vinyl monomers, such as vinyl chloride, vinyl acetate, acrylonitrile, acrylic acid, methacrylic acid, etc. The general reaction is shown in Figure 1.24 where X and Y are different possible pendent groups (H, Cl, CN, etc.) Note that the free radical entity has moved to the end of the new molecule where it will attack another monomer molecule and increase the chain length to two monomer units. As long as the free radical end stays "alive" the polymerization process continues. The process ends via several possible reactions as pointed out earlier.

Ethylene—High Pressure Polymerization

The simplest monomer with a reactive double bond is ethylene (X and Y = H in Figure 1.23). Ethylene, which is normally a gas, is converted at very high pressures (15,000 to 45,000 psi) to a liquid and in this state can be polymerized at 100–300°C using free radical initiation. The initiators, oxygen or peroxides, are injected into the monomer stream just prior to entering a stirred reactor. The liquid monomer containing the polymer in solution is then reduced to atmospheric pressure and the gaseous, unpolymerized monomer is recovered, purified, recompressed and recycled to the reactor. The molten polymer stream is fed directly to an extruder and thence through a die that forms a number of strands that are chopped to yield pellets of uniform size and shape.

Commercial units are large, producing typically 100,000 to 150,000 pounds of polyethylene per hour. With the conversion of monomer to polymer being as low as about 10%, auxiliary hardware must be sized to

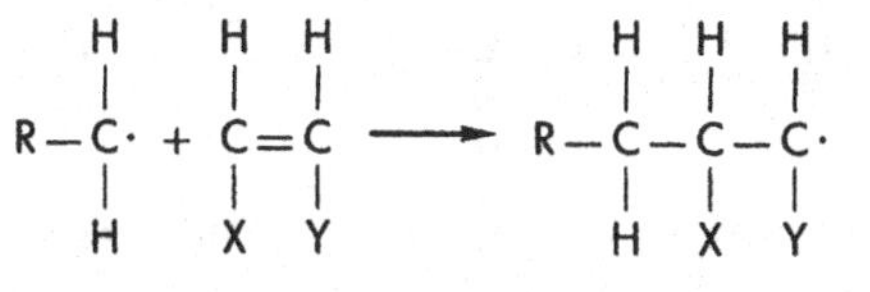

$$
\underset{\underset{H}{|}}{\overset{\overset{H}{|}}{R-C\cdot}} + \underset{\underset{X}{|}}{\overset{\overset{H}{|}}{C}}=\underset{\underset{Y}{|}}{\overset{\overset{H}{|}}{C}} \longrightarrow \underset{\underset{H}{|}}{\overset{\overset{H}{|}}{R-C}}-\underset{\underset{X}{|}}{\overset{\overset{H}{|}}{C}}-\underset{\underset{Y}{|}}{\overset{\overset{H}{|}}{C}}\cdot
$$

Figure 1.24 Free Radical Addition to a Monomer.

recycle large quantities of unconverted ethylene. Because of the high pressure, extensive engineering has gone into the selection of materials and the design for compression, pumping, reacting, separating, purifying, and extruding hardware, with the explosivity of ethylene being a major concern throughout the process. Modern installations make extensive use of computers for control of the process and for handling transitions from one polymer grade to another.

Ethylene Copolymers

Two different monomers will polymerize together to a significant extent if the rates at which they react to free radicals are in the same range. If one monomer reacts very slowly compared to the other, most of the polymerization will take place with the faster reacting one. To some extent, a difference in reaction rates can be offset by increasing the concentration of the slower reacting monomer. Some monomers are not suitable for free radical polymerization because they contain a site that is more chemically reactive than their double bond so that the approaching free radical attacks this site preferentially. An example is propylene, and therefore copolymers of ethylene and propylene are not possible by this route. For the same reason, higher molecular weight analogs of propylene such as butene, pentene, and hexene also do not copolymerize with ethylene in a free radical process.

On the other hand, many monomers that contain oxygen copolymerize readily with ethylene. Examples important to plastic films are given in Figure 1.25. Copolymers of these monomers and ethylene are produced by injecting the second monomer into the ethylene stream. The weight percent of comonomer in the final polymer is usually in the range of 1–20 for

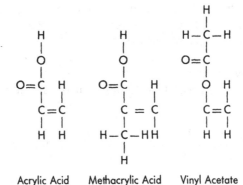

Acrylic Acid   Methacrylic Acid   Vinyl Acetate

Figure 1.25 Monomers That Copolymerize with Ethylene.

copolymers of interest to packaging. A distillation step must be added to the purification process in the unreacted monomer stream, in order to separate the different monomers and insure that a uniform concentration is maintained in the feed stream.

Ionomers are produced by partially neutralizing the acid groups of copolymers made with acrylic or methacrylic acid. This can be accomplished by the addition of sodium hydroxide or zinc chloride in a separate step in an extruder.

Copolymers of ethylene and vinyl alcohol (EVOH) cannot be produced directly since the vinyl alcohol molecule is not chemically stable. Therefore, copolymers of ethylene and vinyl acetate are converted to EVOH in a post-polymerization reaction wherein the acetate groups are hydrolyzed to the alcohol. Because of the high vinyl alcohol content in the copolymers of interest to packaging, the copolymerization of ethylene and vinyl acetate is carried out under much lower pressures than for other ethylene copolymers so that process hardware designed specifically for this purpose is used.

Free Radical Slurry, Dispersion and Bulk Processes

One of the earliest methods developed to synthesize polymers was to stir the monomer and initiator in water. Since the monomers are, in general, immiscible in water, polymerization occurs within the droplets of monomer leading to a slurry or dispersion of solid particles of polymer, usually in very high yield. Thus, the product ends up in a form that is easily pumped and stored and can be used directly in the application such as coating. Another advantage of this system is that the water acts as an excellent reservoir to absorb the heat generated by the polymerization reaction. While the process is usually carried out in a large stirred tank at atmospheric pressure, for monomers that are gases, pressurized tanks are also used. Finally, the process is normally run as a batch reaction although semi-continuous and continuous systems are also used commercially.

Polyvinyl Chloride

The initiator is commonly benzoyl or lauryl peroxide, which is soluble in the monomer but not in water. The suspension of poylmer formed is stabilized with process aids such as methyl cellulose, gelatin, or polyvinyl alcohol, all of which increase the viscosity of the system. Strong agitation is employed and the batch reaction is completed in 6–24 hours after which unpolymerized monomer is stripped out and the polymer particles are filtered, washed, and dried.

A similar approach is to polymerize monomer in bulk (that is, without water). Since the polymer is insoluble in the monomer, it forms as a suspension of particles. Although clarity and color are superior, bulk polymer tends to have a lower thermal stability and a higher molecular weight, both of which make processing into films more difficult.

Polystyrene

Polystyrene is polymerized in bulk, suspension, emulsion, and solution systems. Suspension polymerization, the most common, is run in batch using benzoyl peroxide as the initiator and producing the polymer in the form of clear beads.

Vinylidene Chloride Copolymers

These polymers are produced via the dispersion process using peroxide initiators in both batch and continuous processes. While much of the polymer is sold as the dispersion for coating applications, some is filtered, washed, and dried for solvent coating and for extrusion.

Polymerization by Coordination Catalysis

In coordination catalysis, a monomer molecule is absorbed onto the surface of the solid catalyst and polymerization involves only those molecules that are in this absorbed state. A characteristic of this process is that the catalyst system can be chosen so that the polymer chain is built up in precisely controlled, stereospecific steps. In a stereospecific polymerization each atom on the monomer molecule has the same position as its neighbor along the chain. By contrast, in polymer formed in a free radical reaction, the positions of the atoms on neighboring monomer units tend to be different and in a random pattern. Another characteristic of the structure of such polymers is that branching is minimal, leading to higher crystallinity than is possible for polymers made by the free radical process.

The first coordination catalyst systems (Zeigler catalysts) were combinations of aluminum triethyl with titanium derivatives such as with carbon tetrachloride. The reactive catalyst was a reaction product of this mixture. Later versions include a system of partially reduced chromic oxide supported on steam-activated silica-alumina and nickel oxide on charcoal.

In the earliest versions of this technology, the catalyst is suspended in a solvent in which the polymer is produced as a solution or a slurry. The solvent is then recovered and the catalyst is deactivated and removed from the polymer. More recently, a gas phase process has been developed in which the gaseous monomer passes through a fluidized bed of polymer particles containing the catalyst. A major advantage of all these systems is that they operate at much lower pressures than is required for the free radical system, reducing both the investment for and complexity of the hardware.

Existing coordination catalyst systems work for all simple hydrocarbon monomers (for example, ethylene, propylene, butene, hexene, etc.) but a catalyst system has not yet been found which polymerizes oxygen-containing monomers.

Polyethylene (HDPE, LLDPE, VLDPE and ULDPE)

In one version of the process for HDPE the catalyst is suspended in a liquid hydrocarbon solvent through which ethylene is passed at atmospheric pressure and a temperature of 50–75°C. Granular polyethylene is produced as a slurry that is stirred until viscosity interferes with the dispersion efficiency. The workup of the polymer is as described above and recovered polymer is extruded into pellets as for the LDPE process. In another version, the process is operated at higher pressures (400 to 500 psi) and temperatures (100 to 175°C), producing polymer in a 10% solution in cyclohexane. The gas phase process is also used, in which ethylene, a small amount of hydrogen to modify the molecular weight, and a coordination catalyst are fed to a fluid bed reactor at 280 psi and a temperature of 85 to 100°C.

The copolymers (LLDPE, VLDPE and ULDPE) are made by essentially the same processes as HDPE except for the addition of comonomers.

Polypropylene

The solution process for polypropylene (PP) is similar to that used for HDPE. Pressure is about 1500 psi and the temperature is held at 35 to 40°C, which is low enough to precipitate polymer as a slurry. The catalyst is removed. Recently a gas phase process has been used.

With the stereospecific catalyst system (Zeigler), the polymer has a regular structure (isotactic) with the methyl groups always on the same side: on the outside of the helical chain. Other catalyst systems produce various combinations such as regularly alternating positions of the methyl groups along the chain. Some of the catalyst systems are not stereospecific and rubbery (atactic) polymers are produced, which are of little use in film applications. An atactic fraction is produced in small quantities even with a stereospecific system and must be removed by extraction.

Copolymers with ethylene are made in the same process with ethylene content in the 1 to 8% range.

Condensation Polymerization

In the condensation polymerization reaction, two molecules with reactive sites at both ends react to form two new molecules: one an addition of the two main parts and the second a small molecule made up of fragments split out of the two reactants. An example is the formation of a polyamide from hexamethylene diamine and adipic acid as shown in Figure 1.26.

In general, the condensation reaction can take place under the influence of heat alone; however, in many cases it is a reversible reaction so that as the concentration of by-product increases the reaction begins to get driven back the other way. To achieve high molecular weight polymers, catalysts are used and often the reaction is run under vacuum to remove volatile by-products. As compared to free radical initiated reactions, condensation polymerization requires substantially more energy to drive the reaction.

Nylon

So-called 6,6 nylon is made from adipic acid and hexamethylene diamine as shown in Figure 1.26. Other alternatives include so-called 6 nylon which is made from the single molecule, caprolactam (Figure 1.27), by splitting out water. Nylon copolymers are made from longer chain diamines with adipic acid.

In the process for 6,6 nylon, a salt is made first which is the combination

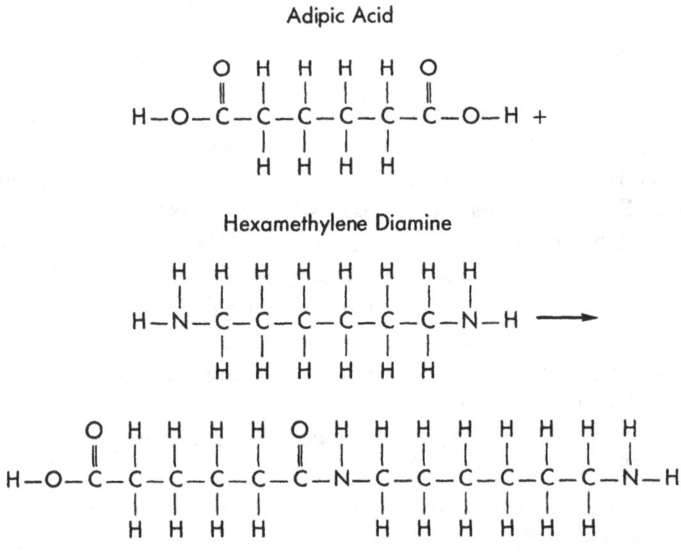

Figure 1.26 The Reaction between Hexamethylene Diamine and Adipic Acid.

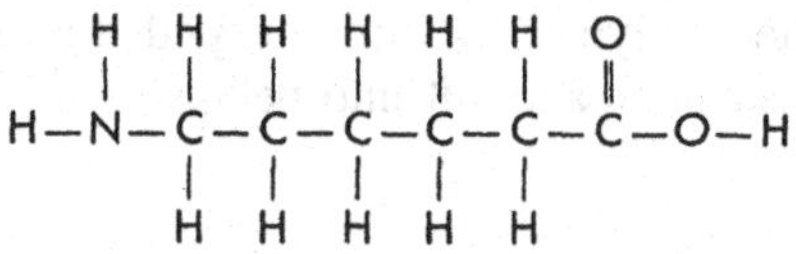

```
      H   H   H   H   H   H   O
      |   |   |   |   |   |   ||
  H—N—C—C—C—C—C—C—O—H
      |   |   |   |   |
      H   H   H   H   H
```

Figure 1.27 Caprolactam.

of one molecule each of the diacid and the diamine. The reaction takes place in boiling methanol with the insoluble salt precipitating out. Then a 60% aqueous solution of the salt with a trace of acetic acid is allowed to react in a stainless steel autoclave under a nitrogen atmosphere at 250 psi pressure. The temperature is staged from 220°C to 270 to 280°C over a period of about 3 hours to achieve high molecular weight. Finally the molten polymer is extruded by nitrogen pressure as a ribbon onto a water cooled wheel.

Polyethylene Terephthalate

In this polymerization, dimethyl terephthalate is reacted with ethylene glycol as shown in Figure 1.28. Sometimes terpthalic acid is used and the by-product is water. In either case, the by-products must be removed, and catalysts are used to drive the reaction to high molecular weight polymers. The process is staged since different catalyst systems are more effective at different stages of the reaction and different levels of vacuum are required. In the first stage, ester exchange, a precursor molecule is produced by the addition of a molecule of glycol to each end of the diacid. This is done at about atmospheric pressure and 150°C using antimony trioxide and cobaltous acetate as a catalyst. In the second stage, this monomer is reacted at 270 to 285°C at a vacuum of 1 mm Hg to form high molecular

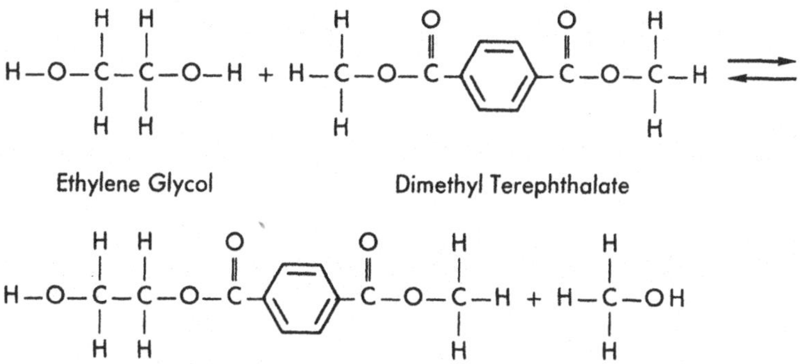

Ethylene Glycol   Dimethyl Terephthalate

Figure 1.28 The Reaction of Ethylene Glycol and Dimethyl Terephthalate.

weight polymer. Molten polymer is then pumped directly to the film-making process or is extruded and cut into pellets.

SUMMARY

Of all these polymerization processes, by far the simplest is when the monomer is a liquid at room temperature and can be readily dispersed in water and initiated via free radicals. Deviations from these conditions quickly lead to increasingly complex hardware and more demanding operating technique. Gaseous monomers require high pressure compressors, piping and reaction vessels. Solid catalysts are very sensitive to poisoning and leave undesirable residues in the resin. Melt condensation polymerization requires very high energy, long reaction times and often large vacuum vessels. As a result, the manufacture of the resins described in this section is the provence of large, technically sophisticated companies.

BIBLIOGRAPHY

Bakker, M., ed. 1986. The *Wiley Encyclopedia of Packaging Technology.* New York, NY: John Wiley and Sons, Inc..

Briston, J. 1983. *Plastics Films. Second Edition.* Essex, England: Longman Group Limited.

Deanin, R. D. 1972. *Polymer Structure, Properties and Applications.* Boston, MA: Cahners Publishing Company, Inc.

Flory, P. J. 1953. *Principles of Polymer Chemistry.* Ithaca, NY: Cornell University Press.

Mandelkern, L. 1972. *An Introduction to Macromolecules.* London, England: The English Universities Press, Ltd.

Sperling, L. H. 1986. *Introduction to Physical Polymer Science.* New York, NY: John Wiley and Sons, Inc.

CHAPTER 2

Processes for Making Plastic Films

INTRODUCTION

The processes for making films that will be discussed in this chapter transform solid polymer resins in granular or pellet form into films on rolls. The first part of the transformation is the extrusion sequence, which consists of the following steps:

Polymer Feeding → Melting → Mixing → Metering → Filtration

All extruded polymers must go through the complete sequence, and variations in each step to accommodate differences in polymer characteristics are relatively subtle. Therefore the description of these steps will be general with an emphasis on fundamentals.

The second part, the film-making sequence, consists of the following steps:

Melt Film Formation → Quenching → Orientation → Windup

The orientation step may or may not be used depending on the polymer. In all of the steps, significant variations are found either because of the characteristics of the polymer involved or because of the demand for specific properties. The approach in this section will be to describe each type of film process, to compare their advantages and disadvantages, and to give process and product details for specific polymer systems.

THE EXTRUSION SEQUENCE

The production of a plastic film that is uniform in width, thickness, and physical properties, and free of bubbles, gel, and undesirable color, requires the extrusion sequence to deliver a clean melt stream at a constant flow rate. A single piece of equipment, called an extruder, performs this sequence. An example is shown in Figure 2.1. At the feed end, solid polymer particles drop onto a screw which rotates inside of a barrel where

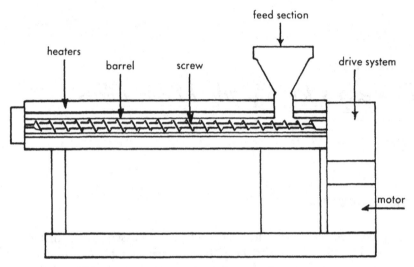

Figure 2.1 A Cross Section of an Extruder.

melting and pumping take place. In the last step, a filter removes foreign matter and gel.

As the variety of polymers and mixtures have steadily increased and the requirements of the film formation sequence have become more demanding over the past thirty years, extruder technology has evolved to the point where screws tailored for specific systems are now readily acquired from the many manufacturers of extruder equipment. In size, extruders are available from bench top models for laboratory work to eighteen inch diameter machines. Typically, commercial film operations use 3-1/2- to 6-inch diameter extruders. The length of the screw is dictated by the diameter and the characteristics of the polymer system. Commonly used length-to-diameter ratios range from 15:1 to 30:1. Materials of construction are usually some form of steel. The screw is usually plated and often barrel liners are used. For chlorine-containing polymers like PVC, evolution of hydrochloric acid from the melt requires special corrosion-resistant surfaces. Another factor of the design, is that the screw is supported and driven from only one end. As a result, the inevitable small amount of wobble in the unsupported end of the screw coupled with the close spacing between the screw and the barrel can cause rapid wear especially as the length-to-diameter of the screw increases. Therefore the easiest surface to replace (the barrel liner) is often designed to be the softer wear surface.

Another important consideration is the drive motor and coupling system for the screw. To deliver a melt stream with the required uniformity of flow, drive speed must be controlled within ±1% or less. Furthermore, the drive motors and couplings used must be large, both to overcome frictional forces and to pump the melt. A common range of motors for film operations is from 50 to 200 horsepower for 3-1/2- to 6-inch extruders.

Polymer Feeding

At the beginning of the extrusion sequence, a feed section in its simplest form consists of a large funnel holding an excess of polymer particles that drop by gravity into the extruder barrel. For modern commercial operations, this section has become more sophisticated in order to insure a constant particle feed rate. Most solid polymers as isolated from the polymerization process are unsuitable for easy handling and uniform flow. In some cases, because the particle size varies over a wide range or because the particle size is very small, clogging or "bridging" of the feed throat of the extruder occurs. Bridging can be overcome by mechanical assists such as a small feed screw in the throat and by vibration of the feed hopper. More often, however, these difficulties are overcome by conversion of the polymer into pellets of uniform size in a separate extrusion step prior to shipment as described in Chapter 1.

Many films are made from polymer blends, which are created by mixing pellets of two or more different polymers. Solid additives are also incorporated by such dry mixing. This process is often conducted as a separate step in batch blenders specifically designed to attain good mixing. However, feed sections are available to accomplish this mixing in-line. One approach is to feed each stream to the hopper via a screw, with the ratio of ingredients controlled by the screw speeds. Since such a screw maintains a constant volume flow rate, rather than a constant mass flow rate, a constant particle size for each stream is essential. For more precise control, gravimetric feed systems, where the streams are controlled by weighing, are preferred.

Another difficulty in feeding is that polymers with low melting points tend either to soften in the feed section or to have a tacky surface, both of which cause bridging in the feed throat. Mechanical assists such as vibration and cooling of the feed section can help cure this problem, as can modification of the particle surface by crystallization or by coating it with a fine powder for better slip.

Finally, polymers must be dry prior to extrusion to avoid the formation of steam and bubbles in the film. In some cases, such as polyesters, the presence of moisture also leads to depolymerization during extrusion and loss of physical properties in the final film. For such polymers, pellets must be dried just prior to extrusion, either in the feed section or in a separate drier. For most polymers, moisture is not usually a problem since the particles or pellets are dried prior to packaging for shipment. Some polymers, such as ionomers, tend to pick up moisture and must be shipped in bags with a high moisture barrier.

Melting

The melting of polymers and its relation to molecular structure was described in Chapter 1. There, melting was assumed to occur under ideal conditions: for example, with the polymer sample at a uniform temperature and with temperature changes occurring very slowly. The conditions in an extruder are far from ideal. Taking the simplest case first, heat is supplied on the outside of the extruder barrel by electrical heaters, hot water or steam. Thus, there is a gradient from some temperature at the wall of the barrel high enough to cause melting to some much lower temperature at the wall of the screw. For example, for LDPE this gradient would be from about 120°C to about 20°C.

If polymer pellets are transported by a screw without any heating, plug flow, where polymer pellets move as if a piston were pushing a plug through a pipe, occurs. With a heated barrel—and assuming that they do not stick to the wall—the pellets would continue to move through the zone between the barrel and screw without changing their position. Those in contact with the barrel wall begin to melt, and at the interface between a pellet and the wall, a liquid contacts a solid surface. The liquid near the barrel wall would increase rapidly in temperature, reaching that of the wall. The rest of the pellet in contact with the wall would reach a high enough temperature to melt slowly, since thermal conductivity for polymers is low. Pellets in contact with those next to the hot wall would similarly increase in temperature, but interpellet spacing adds resistance to the heat flow from pellet to pellet. To melt the pellets next to the coolest point (the screw, in this example), the barrel wall temperature would have to be extremely high and very long times would be required. Under such conditions, the liquid polymer in contact with the wall would likely be exposed to temperatures above its degradation point.

Fortunately, in the real situation, the melt sticks to the wall, and this plus the design of the screw in this section, cause movement and mixing of the melting pellets. To understand this design, the geometry of an extruder screw is shown in Figure 2.2. The screw consists of a helical ribbon wrapped around a cylinder with controlled spacing and angle. In a cross section as shown in Figure 2.2, each pass of the helical ribbon is called a flight and the open space between each flight is called the channel. Important parameters are D, the inside diameter of the barrel, H, the channel depth, and c, the clearance between the top of the screw flight and the barrel surface. E is the pitch and is related to the helix angle, θ. W is the width

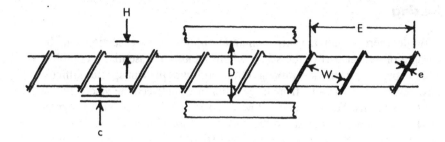

D = inside diameter of the barrel
H = channel depth
c = clearance between screw flight and barrel surface
E = pitch
W = width of channel
e = width of flight

Figure 2.2 Geometry of an Extruder Screw.[1]

of the channel and *e* is the width of the flight. As the screw turns and the channel is moved forward, the melted polymer layer at the barrel wall moves toward and accumulates against the rear flight leading to a circular flow in this area. This flow induces a high degree of mixing of the melted and unmelted polymer, promoting rapid melting. Screws designed for this "heat transfer" melting feature a gradually decreasing depth of the channel moving downstream so that increasing compression deforms and moves the polymer particles as they soften. For polymers with sufficient thermal stability, the flights can be designed to permit a small amount of backflow to enhance the mixing of hot melted polymer with the cooler pellets. For polymers with high melting points, the extruder screw can be heated with a heater in its hollow core. Extruders like the ones described so far melt only by heat from the barrel wall and heat transfer and melt pumping depend roughly on the square of the barrel diameter. Thus, a 4-inch extruder delivers about 16 times the rate of a 1-inch extruder.

Another important assist to melting is frictional heating caused by a screw turning at high speed dragging highly compressed pellets and melt against the barrel wall. Actually, this source of heat is so large that in modern extruders it can be the sole source of heat for melting once the extruder is operating under steady state conditions. In fact, in some cases the barrel must be cooled. Screws designed for frictional melting feature an abrupt reduction in channel depth after only two to three flights.

[1]McKelvey, J. M. 1962. *Polymer Processing*. New York, NY: John Wiley and Sons, Inc., p. 230.

Mixing

While there is not a sharp demarcation, the mixing section can be thought of as that part of the screw where melting is essentially complete and the primary process is mixing. Often additives present special difficulties. One difficulty results from large differences between the viscosities of the two streams. For instance, a low viscosity oil if present in a polymer mixture will tend to migrate to the barrel wall forming a lubricating layer for the polymer melt. This causes loss of frictional forces to enhance mixing. Another example is when pigment particles are added to a polymer stream. There is a tendency for the particles to agglomerate, and very high shear forces are required to break these agglomerates down.

Given the considerations above, the mixing zone of an extruder is designed for very high shear forces by the selection of the design parameters described earlier in Figure 2.2. A relatively high degree of back flow is also desirable. This is achieved not only by the gap between the screw and the barrel but also by constricting the flow downstream to produce a high back pressure. Other ways to improve mixing include more deliberate interruption and/or reversal of the flow of the melt, which can be accomplished by cutting channels in the screw flights or even totally interrupting the flights.

Even greater efficiency of mixing is achieved in extruders that feature intermeshing twin screws with very close tolerances to maximize shear forces. Such extruders are designed also to permit easy variation of the design of the screws so that conditions can be designed which are optimum for a particular polymer system. Sometimes, for example, maximum shear-mixing and backflow are not ideal. The screws for polymers that readily degrade at the melt temperature are designed for minimum residence time, minimum shear working and minimum back flow.

Many modern extruders have injection ports located at various points along the extruder barrel. Thus, the injection point can be chosen so that a liquid can be added after the polymer has been melted and the melt viscosity reduced by temperature and shearing, thereby minimizing the viscosity difference between the polymer melt and the added liquid. A thermally sensitive polymer can be added downstream after the first polymer has already been melted and worked. Such ports can also be used to remove materials from the polymer melt such as steam or volatile degradation products. To accomplish this removal, screw flights are increased in depth and/or lengthened in the vicinity of the port to reduce melt pressure and facilitate release of the volatile material.

One way to reduce the complexity of screw design is to separate some of the mixing from the rest of the extrusion sequence. The polymer and other components can be mixed in a separate extrusion operation set up for the ideal conditions of mixing for that system, to produce pellets that then can be fed to the film extrusion sequence. Sometimes this separate mixing operation is done for the film manufacturer by polymer producers or by shops that specialize in such an operation. As a result, commonly used additives are available commercially as concentrates in a variety of polymer matrices. An example would be LDPE with 1% of white pigment which, when diluted with virgin LDPE as a dry pellet mixture in a 1:10 ratio, leads to an extruded film with 0.1% pigment level. In many cases, such concentrates in one polymer are compatible, at least at low levels, in a range of other polymers. For large volume applications, additive systems can be mixed into polymers at the time the polymer is manufactured.

Finally, although the emphasis in this section has been on mixing to achieve efficient melting and uniformity of the composition of the melt, another requirement is uniformity of the temperature of the melt. Temperature variations in the melt exiting the extruder will persist into the film formation sequence leading to nonuniformities in film thickness, non-flatness of the film, and variations in physical properties of the film especially when film orientation is involved. Therefore, it is often desirable to have a second mixing zone at the end of the extruder screw following the metering section described below.

Metering

Typically, the final section of an extruder screw is dedicated to the precise metering of the melt to the exit of the extruder. For this purpose, screw flights tend to be shallow and clearances between the screw and the barrel wall kept to a minimum. Sometimes this step is divorced from the extruder because of the difficulties of maintaining precise tolerances and drive speeds. In these cases, a melt pump is inserted between the extrusion and the film-forming steps with its own precision drive motor and a pumping mechanism designed solely for that purpose.

Filtration

The purpose of the filtration section is to remove all the foreign matter in the feed stream and the gel produced in the extrusion process. While this fulfills a very necessary function, it also adds difficulties. As matter builds up on the filter medium, the resistance to melt flow increases, causing the pressure to increase since the extruder is maintaining constant mass flow. As the pressure approaches a critically high value, the filter medium deforms and allows gel to pass through it. Also the maximum pressure that the extruder itself can safely tolerate is eventually reached.

To minimize these difficulties, extreme care is taken to avoid all sources of contamination: airborne dirt, cross-contamination of polymers, packaging material, metal from wearing parts in the polymer-conveying system or the extruder, etc. Specifications are established for purchased polymers to assure a tolerable level of gel and contamination.

However, a major source of contamination is the reuse of waste generated in the film making process. Carelessly handled film or trim from slitting operations can pick up dirt. Often the films are combined, in later operations, with other materials that would be incompatible in the melt phase (higher melting polymers, paper, aluminum foil, etc.). These various waste streams can easily get mixed in a film slitting operation if trim is fed to the wrong storage bin.

Once the control of the feed stream has been maximized and the conditions of extrusion are optimized to minimize gel formation, the appropriate filtration technology must be selected. A variety of filter media offer a range of options, balancing the size of particles that will be retained or passed, the capacity to build up particles before reaching critical pressures, and the chemical inertness to a specific polymer melt. Typical filtration media include diatomaceous earths, sintered metals, and metal screens or meshes. All of these materials must be contained in a system that will withstand the force of melt driven by the maximum design pressure of the extrusion system.

The filter medium must be easily replaceable. Replacement can be accomplished by shutting down the extrusion operation, withdrawing the filtration section, and exchanging it with a virgin or cleaned section. Where relatively long times (e.g., days) between filter changes are possible, this is an adequate solution even for commercial operations. However, where more frequent changes are required, the resultant disruption leads to loss of production time and, more importantly, to production of off-specification product made immediately after the operation is started up again. For complex film forming sequences like those including orientation, such interruptions can lead to nonstandard film operating conditions for as much as ten to fifteen minutes. For these situations, filtration systems have been developed that allow the generation of a new filter medium without stopping extruder operation. In one approach a new filter section is inserted as the used one is withdrawn in a synchronized push-pull arrangement. For film operations that are highly sensitive to small variations in melt flow conditions this change can lead to a significant alteration in film properties. Also, there is a tendency for small amounts of material accumulated on the used filter to be dislodged and pass through in the melt. An improvement in this situation is achieved in filters that consist of a continuous mesh or screen that can be moved through the filtration section.

MELT FILM FORMATION

The previous section covered the sequence of steps required to deliver a uniform, clean stream of a polymer melt at a constant flow rate. The next step is to shape that stream into a thin film of molten polymer. A simple way to accomplish this would be to allow the melt to fill a horizontal, dead-ended pipe with a slot cut out along part of its length. A cross section of such a slotted pipe is given below in Figure 2.3. Here, R is the radius of the pipe, Y is the thickness of the pipe wall, and H is the width of the slot. The length of the slot is designated as L. To insure uniform film thickness, it is not enough just to carefully machine the slot to a constant width within the tolerance desired. Fluid dynamics are such that the flow of even a simple, Newtonian liquid like corn syrup is less at the dead end than at the feed end because the pressure decreases from its highest point at the feed end to its lowest point at the dead end. This effect can be described by a

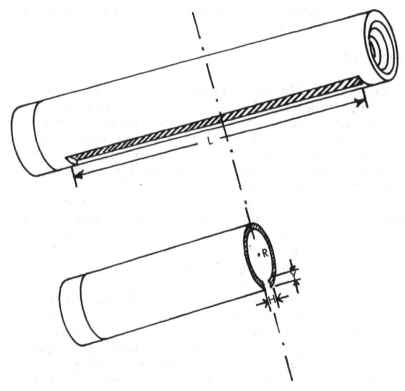

Figure 2.3 Dead Ended Slotted Pipe.

uniformity index, E, which is the ratio of the flow at the dead end to the flow at the feed end of the pipe. For a Newtonian fluid, this is not a large effect and it can be minimized by going to large diameter pipes. For example, where R is 0.35 inches, L is 66 inches, H is 0.010 inches and Y is 0.4 inches, E is 0.92. Increasing R to 0.5 inches raises E to 0.98.[2]

However, polymeric systems do not flow like simple, viscous fluids. As described in Chapter 1, the flow of polymeric systems is non-Newtonian and is governed by a power law [Equation (2)]. As a consequence, the decrease in flow from the feed end to the dead end of a slotted pipe is much larger for polymer melts. For example, for a polymer melt with a flow index of 0.5 [see Equation (2)], E is 0.59 for the slotted pipe above where R is 0.35 and is 0.88 where R is 0.5 inches.[2] Fortunately, E is a more sensitive function of the geometric parameters of the pipe for polymer melts versus Newtonian fluids, so that flow is responsive to relatively small changes in slot opening, pipe radius, and pipe length. These flow relationships can be calculated from complicated but well defined equations provided the rheological data has been obtained for the specific polymer in question. Using this approach, the modern film die has been evolved to minimize differences in flow along the die opening.

Flat Film Dies

An example of a typical film die is shown in Figure 2.4. Note first that the wall of the pipe or manifold is massive so as to achieve mechanical rigidity and to aid thermal uniformity. Second, the manifold is fed into the center rather than the end to reduce the distance over which flow will vary. Also, the manifold is often tapered toward the dead ends to compensate for the falloff in pressure. A third feature is that the slot opening is made variable using a flexible "lip" and adjusting bolts that flex or bend the lips. This offers a final, fine adjustment to reduce variations in film thickness. A selection of designs of adjustable die lip mechanisms is available, each with its own set of advantages and disadvantages. Automatically adjusting die lips are now available commercially in which input from downstream devices that measure film thickness variations is used to control small motors that drive separate adjusting bolts on the die lip. So successful have been these advances in film die design that today film as wide as 100 inches is produced with a uniformity of thickness of ±3% across the width.

The achievement of such uniformity is realized not only by controlling the mechanical precision of the die. Temperature uniformity is also essential. For example, for polyethylene, if the thickness variation is to be ±2.5% across the die, a uniformity index of 0.95 is needed, which trans-

[2]Carley, J. F. 1954. *J. Appl. Physics 25*, p. 1118.

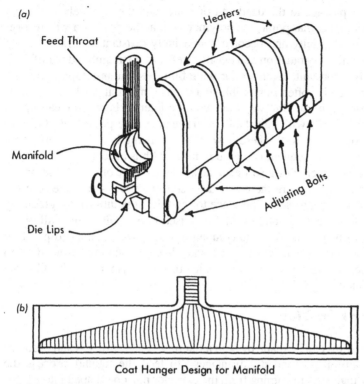

(a)

Feed Throat

Heaters

Manifold

Die Lips

Adjusting Bolts

(b)

Coat Hanger Design for Manifold

Figure 2.4 Center Fed Film Die: (a) Cross-sectioned Perpendicular to the Manifold at the Feed Throat, (b) Cross-sectioned Parallel to Manifold Showing Coat Hanger Manifold.

lates to a temperature uniformity from feed end to dead end of ±1°C. Typically, the film die is heated by a series of electric heaters or by fluid passing through channels in the die. Thermocouples placed at intervals near the die lips are used to monitor the temperature uniformity. Usually, the die body is well insulated to conserve heat and to reduce the effect of ambient temperature fluctuations.

Finally, another important consideration in achieving uniformity in the melt film is thermal degradation. The residence time for the melt at the dead end of the die can be longer than for the melt near the feed opening. Thus, the polymer can degrade at the dead end, building up gel and other degradation products. The die opening near the dead end can gradually become plugged over time.

The situation is exacerbated in polymer melts that are strongly viscoelastic. As described in Chapter 1, the elasticity of such melts is gradually reduced by shear working to a point where flow resembles a normal, viscous melt, but it is recovered again on resting. Given the long time required for passage to the dead end of a die, and the relatively low shear rates obtained at that point, the polymer melt at the dead end will recover some of its elasticity and become increasingly resistant to flow.

To minimize degradation and elasticity effects, a streamlined manifold is extremely important. Even so, for wide film dies, the buildup of polymer at the dead end is often impossible to avoid and necessitates frequent shutdowns for clean up. One way to extend the interval between clean ups is to use end blocks with adjustable inserts that can be pushed in gradually to block off the sections where flow control is being lost. This works only so long as the consequent decrease in film width is acceptable. Buildup of degraded polymer will also occur at any other part of a film die or the system feeding it where there is a dead spot with low shear rate or a hot spot. Thus it is common experience to see thickness uniformity gradually deteriorate as an originally clean die is used to the point where all of the adjustment mechanisms fail to compensate adequately. This end point, at which the die must be dismantled and cleaned, can vary from days to weeks depending on the characteristics of the polymer and the film die system itself.

Tubular Film Dies

Plastic films can be made by both a flat or a tubular process and each has its own advantages and disadvantages. To feed the tubular process the polymer melt stream coming from the extruder must be shaped into a tube. While the principles for flow that dictate the design of a tubular die are much the same as for the flat die described above, there are some differences. A typical tubular die is shown in Figure 2.5.

In this die the polymer melt flows through an annular channel that is formed by the outer die body and the inner die mandrel. The mandrel is supported mechanically by "spider legs" attached to the die body. A problem with this design is that the melt forms "weld lines" as the flow is split and reforms around the spider legs. These weld lines will not only be visible in the final film but will constitute variations in film thickness. Spider legs can be eliminated using a cross-head design in which the mandrel is attached at its base to the die body and inlets for the melt are just above this base perpendicular to the annular flow channel. Because the flow of melt is split by the mandrel, some weld lines are also present for the cross-head die and flow nonuniformities due to the different lengths of the flow path around the mandrel, and are difficult to compensate for by adjustments in the die geometry. These defects are largely eliminated for the annular or cross-head designs by machining spirals into the mandrel between the inlets and the die lips that disrupt and divert the flow to cause sufficient intermixing of the streams for them to lose their identity. This mixing also

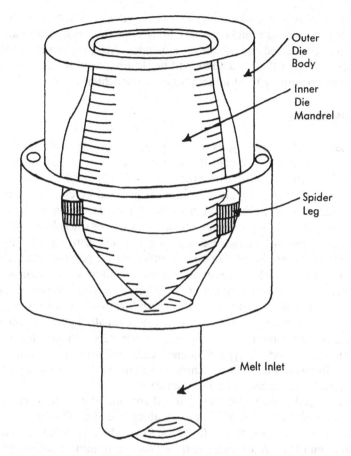

Outer
Die
Body

Inner
Die
Mandrel

Spider
Leg

Melt Inlet

Figure 2.5 Tubular Die Design.

eliminates stratified temperature variations set up in the melt as it flows from the extruder through the connecting pipe to the die.

Unlike flat film dies, the lips of tubular film dies cannot be flexed for improved thickness control. As a result, thickness uniformity in tubular dies is ±10% versus ±3% for flat film dies.

CAST FILM PROCESS

By means of the flat or circular film dies described above, the polymer melt can be shaped into a thin film or tube. Next the film must be solidified by cooling or quenching and then wound up on a roll. These steps will be described next, first for the cast film process and then for the tubular or blown film process.

As the melt film emerges from the die opening it is critical to quench the film quickly to minimize thickness variations since the weak melt tends to draw in an uncontrolled and nonuniform manner. Also, the more slowly a melt cools the larger the crystallites will be, leading to film haze. Several methods are commonly used to achieve high quenching rates.

Roll Quenching

One method of quenching is to bring the emerging melt film immediately in contact with a cold metal surface using a rotating drum or quench roll. In practice the distance between the die lips and the drum surface is between 1 and 2.5 inches. Cooling is attained by circulating cold water or antifreeze solutions through the hollow drum. Often the surface is chromium plated to improve heat transfer and insure smoothness. If the surface is cold enough to condense moisture droplets, care must be taken to remove these to avoid a pattern on the surface of the polymer film. This need for a clean roll surface also applies to other sources of contamination such as dirt, material migrating out of the polymer melt and condensing on the drum, etc. For films that require a very uniform surface, extreme care must be taken to clean the roll surface either periodically by shutting down the process or by a continuous roll cleaning system such as a rotating brush with vacuuum take-off. A typical quench roll configuration is shown in Figure 2.6. In this example, an air knife is used to provide additional force to achieve intimate contact with the quench roll.

The quench roll must be precisely aligned with the film die to insure that draw tensions are constant at all points along the die. This minimizes thickness variations across the width of the film, which is called the transverse direction (TD). Also, care must be taken to minimize variations in the quench roll rotation speed or mechanical vibrations in the quench roll system which will lead to periodic variations in film thickness along its length or machine direction (MD). By determining the degree of thickness variation and analyzing this along both TD and MD, it is possible to pinpoint where the source of the variation is likely to be, but there are complications. Any pulsing in the stream as fed from the extruder will also cause MD variations. Given the way the melt flows in flat film die, these pulses would also generate a TD component to the thickness variation. Hence, elimination of potential causes by precise mechanical specifications and operating procedures is essential to minimizing the problem.

The advantage of the quench roll process is that it can operate at high film speeds. As the film speed is increased the roll diameter is increased to provide more quenching residence time and/or the cooling temperature is decreased. In addition to the need for excellent mechanical precision, the main disadvantage is that the melt still cools at a relatively slow rate,

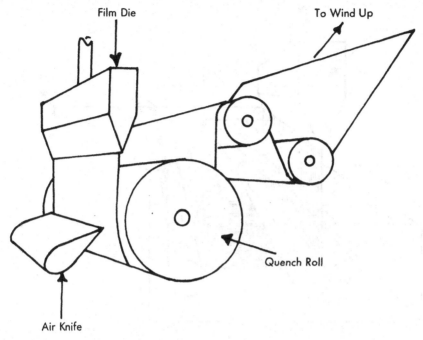

Figure 2.6 Roll Quenching.

partly because only one side of the film is in contact with the cool surface. Cooling rate can be increased by providing a second quench roll for the second side, with the consequent disadvantage of increasing mechanical complexity. A second problem with the quench roll system is that heat conduction across the film/metal interface is inhibited by the presence of a thin layer of air trapped between the film and the metal. Elaborate schemes have been developed to remove this layer and significantly increase the quench rate. Even under the best conditions, however, quenching on a roll still leads to a cooling rate that is slow enough to generate some level of large crystallites and some undesirable cloudiness or haze in the final film for many crystalline polymers. The clearest films for these polymers are obtained using an alternative quenching method: immediate immersion of the film in a bath of water.

Water Quenching

A diagram of a typical water quench system is shown in Figure 2.7. Water quenching provides the ideal conditions for rapid cooling of the melt: elimination of the air layer, two-side contact, a cooling medium of

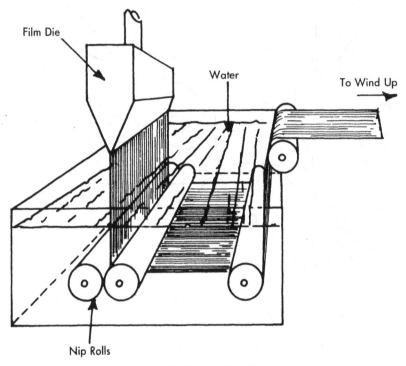

Figure 2.7 Water Quenching.

high thermal conductivity, and, with adequate stirring of the water, a constant cooling temperature. Turbulence of the water surface must be minimized to avoid any patterns forming on the surface of the film. Also, the water droplets carried along by the film must be removed, usually by a simple squeegee blade. Thus the major disadvantages of the quench roll system are overcome, but, mechanical imprecision in the alignment of the take-off roll or in the take-off speed must be minimized just as for the quench roll system in order to minimize thickness variation in the final film, since all of these fluctuations get translated directly back through the solidified film to the weak melt film. The major disadvantage for the water quench system is that film speed is limited by turbulence and slinging of the water bath, which occurs at speeds below the maximum for roll quenching.

As might be expected, various combinations of quench roll and water quench systems are used, as are systems that employ thin layers of water carried along by the moving film and squeegeed off downstream. Since the quenching system is often the bottleneck as film speeds are pushed to higher and higher levels, it remains an area of continuing development.

Film Edge Effects

One characteristic of the cast film process is that the maximum final film width is limited to the width of film die. Where wide widths (for example, 100 inches) are desired, wide film dies, with all of the complications described above, must be used.

A second characteristic is that a thickened edge is unavoidable. This results from the tendency of the melt, in the air gap between the die and the first contact with the quenching surface, to "neck-in" and thicken at the edges due to the viscoelastic nature of the melt. The amount of neck-in varies greatly from polymer to polymer and can be altered somewhat for a given polymer by changing the molecular structure. It can also be lessened by minimizing the length of the air gap. But in the end, the thickened edges must be slit off, usually before the film is wound on a roll. If they are not, they build up on the winding roll, creating a "dished" roll leading to distorted film and a loss in sheet flatness. Also, the built-up edges often create an unstable roll with a nonuniform edge so that the film weaves back and forth on the roll.

Slitting off the thickened edge leads to a loss in the film width of from 1/2 to 2 inches. Since this trim is usually captured and recycled back to the extruder, it does not represent a material loss but only a decrease in productivity and maximum available width.

Measurement of Film Thickness

Given the dynamic nature of the film process and the complexity of the variables that can affect film thickness uniformity, it is essential to measure film thickness continuously as part of the film formation sequence. This is normally accomplished by a radiation device (see Figure 2.8) located upstream of the winding roll. The amount of radiation absorbed depends strongly on the chemical structure of a film. While one film may be transparent to a specific kind of radiation another may completely absorb it. Complete absorption is undesirable since variations in thickness would not be seen. Therefore, the radiation source is chosen to achieve an intermediate level of absorption by the film. A common source is beta-radiation produced by radioactive materials. X-rays and infrared and visible radiation are also used. In some cases, for a given source the range of wave lengths used can be selected to maximize absorption by the film.

Given the correct choice of the radiation source, variations in the amount of radiation absorbed accurately reflect variations in film thickness. If the device is kept stationary, then the variations are being measured in the MD as the film passes by. If the device is moved across the film on the track shown in Figure 2.8, then the differences are caused

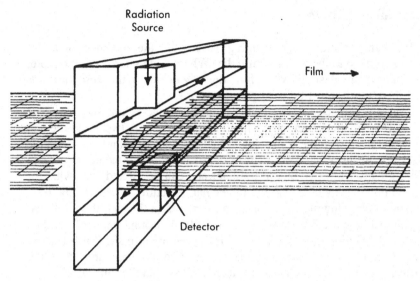

Figure 2.8 Radiation Device for Measuring Film Thickness.

by variations in both the TD and MD since the film is moving in the MD during the TD trace. Repeat scans in the TD are used to identify the TD variations alone since these persist while MD variations change. In modern systems, process computers readily transform these data into pictures of the MD and TD variations.

A typical trace for a plastic film is shown in Figure 2.9. Usually, average thickness variation (see Figure 2.9) is used as a measure of film thickness uniformity. Localized variations cause problems in film winding. For example, sharp spikes in the trace are called gauge bands and account for major problems that are described in the next section. A wedge-shaped variation, typically near the edge of the film, is referred to as "tapered gauge". This defect causes a tapered roll of film, which is unstable, and film tends to slip off the end of a roll in a telescoping fashion.

Film Winding

The process of winding the finished film has its own set of complications. A simple method is "center winding" where a core is driven at a constant rate such that its surface speed is initially the same as that of the film. As the roll diameter increases so does the surface speed. This in turn increases film tension, which can cause several problems. First, as the roll is more and more tightly wound it will tend to cinch non-uniformly leading to buckles or folded waves of film being formed near the core, as shown

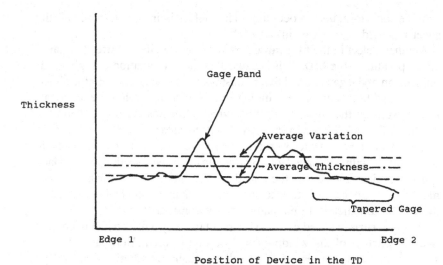

Figure 2.9 Typical Trace of Film Thickness Variations.

in Figure 2.10. This problem is amplified by the presence of air, which is carried into the roll as a tightly adhering surface layer, resulting in layers of air between the film layers. As the tension increases on the outer winding layers this air is squeezed out of the inner layers causing the inner part of the roll to collapse. With very high tension the film even tends to stretch, usually within the elastic limit of elongation. Upon storage the stretched film within the roll tries to recover this elastic deformation causing further film movement on the roll and even greater distortion or cinching. When such a roll is unwound for use, the film in the buckled regions has stretched and deformed to the extent that it will not lay flat and is un-

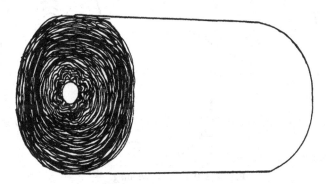

Figure 2.10 Film Roll with Buckles.

suitable for subsequent processing. This defect is in a pattern across the sheet repeated at intervals in the MD.

Another defect in the film caused by winding is called a "stretched lane" and is parallel to the MD. This is caused by the combination of high winding tension and a gauge band that persists in the same location in the sheet. Such a band builds up in the winding roll, increasing even more the localized tension on the surrounding film which stretches the film beyond the elastic limit, causing a permanently distorted area.

One of the best ways to overcome these problems is to have a variable speed drive with the speed of rotation programmed so that the surface speed remains constant as the roll diameter increases. A further refinement on this approach is to measure film tension and control the speed of the core automatically to maintain tension constant.

Constant surface speed can also be maintained by a driven roll in contact with the surface of the winding roll (see Figure 2.11). Here also the film speed is kept constant regardless of roll diameter. An additional advantage of surface winding is that the downward pressure of the driven roll on the film tends to squeeze out the air layer creating a film roll that is more stable in storage. On the other hand, this roll is more tightly wound than a center wound roll and can yield more exaggerated stretched lanes and other distortions. It is difficult to produce the ideal conditions in all cases by either

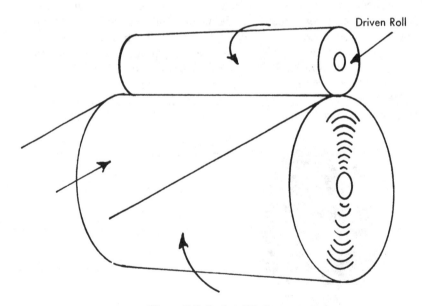

Figure 2.11 Surface Winding.

center or surface winding, and therefore combinations of surface and center winding are sometimes used for the optimum control. Often a non-driven roll is used on a center driven wind up to squeeze out the air layer.

However, even with the best mechanical equipment and control of the winding operation, defects due to small gauge bands are generally unavoidable, especially in large diameter rolls. Randomization of the position of the gauge band in the winding roll, which avoids the buildup of the band, is a valuable assist. A common way to do this is to place the winding system on a platform that oscillates back and forth. The oscillation amplitude and speed need only to be a few per cent of the film width and speed respectively to distribute the gauge bands adequately. To maintain a uniform edge on the film roll, additional trim must be taken essentially equal to the oscillation amplitude.

Film Surface Effects

Typically, plastic films have such smooth surfaces and such intimate layer-to-layer contact that a high force is required to move one film layer against another. Instead of film surfaces sliding past one another smoothly and uniformally, there is a "stick-slip" movement which results in a pattern of soft and hard wound areas in a roll. In addition, the smooth surfaces are cushioned even more effectively by the trapped air layers causing uncontrollable telescoping. These difficulties are overcome by roughening the surface of the film. Imprinting a pattern on the surface during the quenching process or inducing some crystallization helps, but both of these approaches tend to produce undesirable surface haze. Therefore, the more common approach is to add hard particles such as silica or a high melting polymer usually at a concentration of 1% or less. Sometimes, a match of the indices of refraction of the polymer and the particles can be made to minimize light scattering. Otherwise, a balance must be sought between film clarity and surface slip. Since the larger particles (1 to 10 microns) are the most effective in improving slip while the smaller particles only produce haze, it is best to use a narrow distribution of larger particle sizes. Another additive system often used is a waxy substance that migrates to the surface. Again, the balance of film clarity and surface slip must be sought.

BLOWN FILM PROCESS

In the blown film process a tube is formed and then, while it is molten, is blown up like a bubble to generate a large diameter tube from a relatively small circular die. A typical system is shown in Figure 2.12.

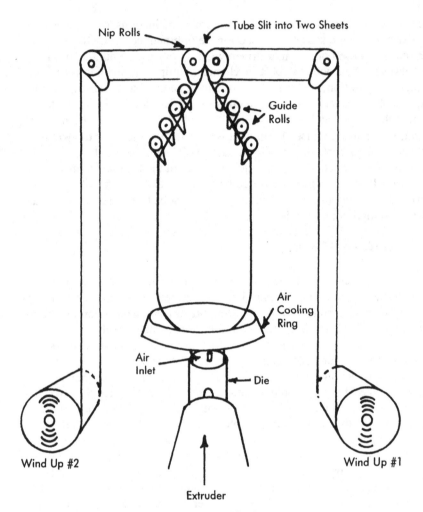

Figure 2.12 Blown Film Process.

Blowing and Air Quenching

As the molten polymer film emerges from the circular die it is immediately inflated by air pressure within the tube. One means of accomplishing this is to inject air through a tube within the core of the die. Downstream a wide nip roll pinches off the inflated tube to maintain the air pressure. A second method is to inject the air via a tube that fits in a groove between the nip rolls. The air pressure is typically quite small and must be controlled with some precision since the molten film is weak.

Downstream of the initial blow point for the bubble, external air is blown against the outside of the bubble by a circular vent to begin to control the rate of expansion by rapidly cooling the film. Since the bubble grows during inflation in both directions (TD and MD), the film speed at the downstream nip rolls must be faster than the tube speed as it emerges from the die. Typical MD stretch (ratio of initial to final film speed) and TD blow-up ratios (ratio of final to initial tube diameter) are in the 2:1 to 5:1 range.

The major strength of the blown film process can be seen in the following example. At a 5:1 TD blow-up ratio, a relatively small, 12-inch diameter circular die will produce a flattened tube 90 inches wide. Thus two sheets of 90-inch width are produced by slitting this tube at each edge, as compared to a single sheet produced by a massive flat film die 91 or 92 inches wide in the cast film process. Furthermore, given an additional 5:1 MD stretch ratio, the total thickness reduction would be 25:1. Thus a die opening of about 25 mils would yield a 1 mil film. This larger opening greatly simplifies the extrusion process because of the greatly reduced resistance to flow and correspondingly low extrusion pressures. The weakness in the blown film process lies in the difficulty in achieving uniformity of the melt drawing process that occurs during the blowing.

Melt Drawing

Assuming that the air pressure and, therefore, the drawing force is constant at all points in the bubble, the amount of draw that occurs in any small, local area depends on the temperature, thickness and the viscosity of the melt in that area. A section that is hotter or thinner or that has lower viscosity will draw at an ever faster rate, becoming thinner and thinner. This leads to a magnification of any variations in the cast tube. Ultimately there is some counterbalancing, since thin sections cool faster than thick sections. Thus, efficient air cooling of the bubble is very important, as is uniformity of the cooling air around the tube. To eliminate the influence of ambient air currents, the blowing zone is often enclosed in a compartment or by curtains. Even under the best of conditions, however, thickness uniformity for blown films is about ±10% compared to ±5% (or lower) in cast films.

Tube Collapsing, Slitting and Winding

The blown tube is contained between the die and a wide set of nip rolls. The collapse of the tube at the nip point must be controlled to avoid bunching in the nip and to avoid discontinuities in the pulling force being translated back to the drawing melt. This is accomplished either by guide rolls or two smooth boards placed in a V shape with the open end a little larger than the blown tube diameter and the narrow end just above the nip rolls. These boards are sometimes perforated to allow the passage of air, which provides a supporting air layer for the tube.

The tube is usually slit in line. One approach is to insert razor blades into the folds of the collapsed tube just downstream of the nip rolls. For films with low tear strength, small edge wrinkles pressed in by the nip roll can cause a nicked slit edge and tearing across the sheet. Thus an alternative approach is to slit just upstream of the nip rolls. This opens the inflated tube into two sheets that pass through the nip rolls. Note that there is no trim to recycle back to the extrusion sequence at this point as there is for the heavy edge in a cast film.

The two slit sheets emerging from the vicinity of the nip rolls are transported by separate systems of rolls to two different winding stations. All of the principles for producing uniform rolls of film described above for the cast film process apply here as well. However, there is an advantage to the tubular process. Since the process is centered about an axis, any or all components can be rotated or oscillated to randomize variations in film thickness that might be caused by that particular component: the die, the air ring, and the collapser. While this is sometimes done, it is mechanically and electrically complicated. It is difficult to connect fixed electrical, air, and water inlets to a rotating die. The moving parts of the die must be equipped with carefully designed seals to eliminate leakage of polymer melt. Therefore, rotation is more often reserved for the end of the process. This randomizes all the thickness variations that have accumulated throughout the system. Typically, the film-winding stations, and sometimes the collapser as well, are supported on platforms that rotate or oscillate.

CAST AND BLOWN FILM COMPARISON

In this section the cast and blown film processes will be compared by looking at process-dependent properties and the range of application. In general, the choice of one process over another is dictated primarily by the properties desired since in all other respects the differences between them are small.

Process Dependent Properties

The film properties controlled primarily by process variables are clarity, thickness and thickness uniformity, and sheet flatness. The polymer system determines to a large extent the other important properties such as heat sealability, gas barrier, toughness, stiffness and tear strength. Surface topography, which controls film slip, depends largely on additives. The comparison of process-dependent properties will be keyed to the process steps where significant differences exist.

Quenching

Since in a blown film process the melt is allowed to cool over a relatively long period of time, blown films are more crystalline and have a higher haze level than cast films. The difference in haze level in the two processes increases with increasing film thickness. Also, more efficient quenching gives the cast film process a higher throughput for thicker films. Nevertheless, there are some benefits of the higher crystallinity for blown films, such as greater stiffness and improved barrier to moisture and gases. Similarly, cast film quenched on a roll is slightly better in these properties than water-quenched film.

Melt Drawing

Another difference between these processes is the much higher level of melt drawing for blown film, which causes an inferiority in level of uniformity of film thickness and sheet flatness. On the plus side, however, there is a small increase in the orientation of the polymer molecules in blown film primarily in the TD. This increases strength, toughness, and stiffness. Tear strength tends to be higher in the MD and reduced in the TD for blown films. At high melt draw ratios, similar effects of melt orientation occur in cast films but at lower levels.

Summary

This comparison is summarized in Table 2.1.

Range of Application

A major advantage of the cast film process is its broad applicability. Any polymer that can be extruded into a uniform melt stream can be converted

Table 2.1. Comparison of cast and blown film processes.

Property	Cast	Blown
Film Clarity	+	−
Stiffness	−	+
Toughness	−	+
Barrier	−	+
Thickness uniformity	+	−
Film Flatness	+	−
Throughput (thick films)	+	−

into a film by this process. The films produced may not always be commercially useful. For example, some polymers yield brittle films, and special orientation processes are required to enhance toughness. Also, whether oriented or not, a threshold molecular weight must be reached to achieve useful strength and toughness.

The range of polymers that can be converted into films via the blown film process is more limited because the high degree of melt drawing inherent in the process requires a high level of melt strength. Generally, the melt properties for the polymer systems of major commercial interest have been modified so that grades are now available for blown films. Given the lower levels of thickness uniformity, sheet flatness, and clarity, the blown film process tends to be used for applications where somewhat lower quality product can be tolerated. Also, since it is relatively less complicated, it is usually the process of choice in situations where technical support is minimum.

CAST AND BLOWN FILMS—TYPICAL SYSTEMS

In this section only a few film properties are highlighted. For a more complete summary of properties the reader is referred to Table A.1 in the Appendix.

Polyethylene

All of the different types of polyethylene film are made on both cast and blown film lines, but blown film dominates. To give an understanding of typical process conditions and properties, blown HDPE films will be described in detail.

For the blown film process, 1-1/2- to 6-inch diameter extruders are employed with L/D ratios from 25:1 to 30:1 for normal molecular weight polymers. For the high molecular weight versions, shorter screws (18:1 to 21:1) must be used with high screw speeds to minimize residence time. One complete line recently announced uses a 60 mm, 24:1 L/D air-cooled extruder. Making 0.7 mil film from a mix of 90% HDPE and 10% LLDPE, a 9-inch diameter die produces a 60-inch lay flat tube at 450 pph optimum rate. While the line can be run at 600 pph, unacceptable film was produced.[3]

Blown film properties are controlled primarily by crystallinity and molecular weight. As density is increased from 0.935 to 0.965, tensile

---

[3]1989. Plastics Technology July, p. 73.

strength increases from 2600 psi to 5300 psi and water vapor transmission rate (WVTR) drops from 0.66 to 0.24 g-mil/100 in^2-day. Toughness and impact strength are much higher in the high molecular grades now available. The effect of increased melt orientation is seen in one example where impact strength was increased 50% when MD melt draw or TD blow-up ratio was doubled.

Polypropylene

Polypropylene films are generally made by the cast film sequence using chill roll quenching because of the tendency for the polymer to crystallize readily. Success in the blown film process can only be achieved by using water cooling. In this process, the tube, cast downward, is enveloped in a sheath of water cascading downward. Also, the collapsed tube is passed into a quench tank of water. As a result of this high rate of cooling the formation of large crystallites is prevented. By this approach film clarity and strength are about 10 to 20% improved over cast film. While either process successfully inhibits crystallization initially, some does develop over time as exhibited by the modulus increasing about 50% in 21 days. During this same period slip additives also migrate to the surface, reducing the coefficient of friction (COF) to the desired range. For extrusion a high L/D extruder is used (L/D = 28 to 32:1). For the blown film process, blow-up ratios are less than 2:1.

Both homopolymer and copolymers with 1.4 to 3% ethylene are used in cast film. For blown films polymers with higher than normal atactic content are preferred to broaden and lower the temperature range for melt drawing. Ethylene copolymers yield films with lower haze (2% versus 3.5% for homopolymer) but with reduced modulus (85,000 versus 110,000 psi for homopolymer) and service temperature (110° versus 138°C upper temperature).[4]

Polyvinyl Chloride

PVC packaging films are generally made by the blown film process. Because of the thermal instability and the high melt viscosity of the polymer, die heads must be carefully designed to eliminate dead spots. Hence the preference is for a streamlined, tubular die in line with a horizontal extruder, bubble, and take off. This arrangement avoids the additional piping and residence time due to the transition from a horizontal extruder to a vertical bubble and take off in the normal blown film line. As

[4]Bakker, M., ed. 1986. *The Wiley Encyclopedia of Packaging Technology.* New York, NY: John Wiley and Sons, Inc., p. 314.

an example of this process, a 4-1/2-inch extruder produces film at about 700 pounds per hour in a thickness range of 0.5 to 3 mils.[5]

PVC film properties are determined largely by the level of plasticizer used. These are typically di(2-ethyl hexyl)phthallate or di(2-ethyl hexyl)adipate. Fairly stiff films for overwrap contain about 10% plasticizer. Flexible meat wrap films contain about 20 to 25%. Very high levels are used to achieve low temperature performance: 30 to 40% is typical for frozen food packaging. Over this range of plasticizer levels, other properties determined by crystallinity vary as expected. For example, barrier to water vapor increases about 50% as the plasticizer level is reduced from the highest to the lowest.[6]

Nylon

Nylon resins are made into films by both the cast or blown film processes. Rapidly quenched cast films are more amorphous and clearer. More crystalline blown films are hazier but have superior gas barrier.

For successful extrusion, polymer moisture content must be kept below 0.1%. Conventional polyethylene extruders can be used but heating capacity must be increased to accommodate the higher melting point of nylon and improved temperature control ($\pm 3°C$) is required to minimize degradation. There are special screws designed for nylon extrusion that are more ideal than standard polyethylene screws. L/D ratio is in the range of 20 to 24:1.

In the blown film process the stiffness of nylon film causes wrinkling problems in the collapser. Careful mechanical alignment, best thickness uniformity, and reduced surface slip to reduce drag forces all help to minimize the problem.[7]

Polyvinylidene Chloride

PVDC films are made by both cast and blown film processes with the latter used primarily for oriented films. In the cast film process, rapid quenching inhibits crystallization. In general, PVDC films tend to cling and are difficult to wind into uniform rolls. Their outstanding property is their high barrier to water vapor and gases.

---

[5]Benning, C. J. 1983. *Plastics Films for Packaging*. Lancaster, PA: Technomic Publishing Co., Inc., p. 66.
[6]Bakker, p. 310.
[7]Ibid. p. 479.

Ethylene/Vinyl Acetate

Both cast and blown film processes are used. Cast films have better clarity while blown films are tougher and frequently used in shrink bag applications. Because EVA films have greater elasticity than polyethylene films, more careful control of tension throughout the process must be maintained.

At low levels of vinyl acetate (7–8%), EVA copolymers act like modified polyethylene whereas at higher levels (15–20%) films are more like plasticized PVC. Thus the comonomer acts like an internal plasticizer.

FILM ORIENTATION

In the cast and blown film sequences the major contribution of the process to properties, beyond shaping the polymer into a film and winding it on a roll with acceptable control of thickness variation and sheet flatness, is to offer some variation in the level of crystallinity. Another important dimension is added when the process induces significant orientation of the polymer molecules. Orientation brings out the maximum strength and stiffness inherent in the polymer system. In addition, the orientation induces even higher levels of crystallinity so that properties like barrier and chemical inertness are further enhanced. Optical properties are at the same time generally superior, since orientation leads to a crystalline structure that scatters much less light than the crystalline domains formed in unoriented films.

Orientation of Polymer Molecules

As described earlier, in both the cast and blown film processes drawing of the polymer melt causes some small amount of orientation of the polymer molecules. The amount is small because of competing molecular processes. While melt drawing tends to straighten out the polymer molecules and align them in the direction of the force applied, there is always a counter force within the molecule to return it to its natural coiled state. How fast this relaxation occurs depends on the viscosity of the melt, but it is generally fast enough so that very little permanent molecular alignment can be achieved by drawing a polymer melt. An exception is some liquid crystal polymers, which are rigid rods rather than random coils, so that the molecules readily align themselves in the melt.

One might think that orientation efficiency in melt drawing would increase if the melt were allowed to begin to solidify, thus retarding the relaxation process. The problem here is that the crystallization process sets in at a rapid rate, anchoring the molecules into a network before they can be stretched out and aligned. In fact, it has been shown that crystallites do not completely melt or totally lose their identity in the melted state and these nuclei are waiting to initiate the reformation of more crystallites. Furthermore, as the first stages of alignment occur crystallization is accelerated. Thus, in real systems, it is not practical to find just the right point in the time/temperature map for the orientation of a drawn, cooling melt.

The solution is to approach the orientation from the other direction: that is, by heating up an amorphous film. Many crystalline polymers can be formed in an amorphous state by very rapid cooling of the melt. This is not a thermodynamically stable state, so the molecules will crystallize if the solid is heated above the glass transition temperature (see Chapter 1). However, if an amorphous film is heated above the T_g and stretched quickly, the molecular alignment competes favorably with crystallization and the drawn polymer molecules condense into a crystalline network with the crystallites aligned in the direction of the drawing force.

When a film is drawn in only one direction the crystallites align themselves in that direction. The film then exhibits great strength and stiffness in that direction but the strength in the other direction is low. In fact, such films usually fibrillate or tear into fibers readily. While some one-way stretched films are used in packaging, they are not common and the discussion from here on will be confined to two-way or biaxially oriented films. For such a film, the crystallites tend to be sheet-like and the macrostructure of the film resembles a book. When this lamellar structure is strongly developed, as in polyester films, the physical properties are dramatically affected. Evidence of this is seen in the lower strength in the film thickness direction. Thus, failure of the bond to a strongly-adhering coating is often at the interface between the lamellae in the film. Such a film exhibits high resistance to being punctured, even at very high impact forces and speeds, but forces that flex the film cause it to come apart at the weak lamellar interfaces. Biaxially oriented films with less well-defined lamellar structures are less stiff and strong but resist flexing forces better. The tear strength of oriented films is always much lower than that of unoriented films.

In addition to defining the shape and alignment of the crystallites, the orientation process accelerates and increases the extent of crystallization. However, there still remain in the final film some disordered or amorphous regions. Molecules in these areas are anchored in a more straightened configuration as the surrounding crystallites are formed and film temperature reduced. However, if the film is heated to near the second order transition temperature, mobility is restored to these taut molecules and they relax back toward the random coil configuration. The physical manifestation of this is that an oriented film will shrink at temperatures exceeding the second order transition temperature by 5% or more. Therefore, another step must be introduced into the sequence: heat stabilization or heat setting.

The sequence for a film orientation process then is: film formation, quenching, reheating to the orientation temperature, film stretching, film stabilizing, and winding. Either a flat or tubular process can be used to accomplish this sequence.

FLAT FILM ORIENTATION PROCESS

A diagram of a flat film orientation process is shown in Figure 2.13.

Melt Film Formation

This step is essentially identical to film formation in the cast film process. Since TD stretching greatly increases the film width, the casting die width need only be in the range of 1/4 to 1/8 of the final film width. On the other hand, the requirements for uniformity of the molten film (viscosity, temperature, composition, and thickness) are more stringent.

Melt Film Quenching

Again, this step is essentially identical to the cast film process. Usually, a quenching roll is employed (see Figure 2.13) whose width need only be

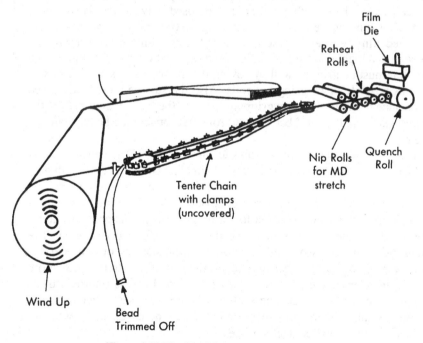

Figure 2.13 The Flat Film Orientation Process.

a fraction of the final film width. Also, as above, better mechanical and operational control over the quenching process must be maintained.

Film Reheating

This, typically, is done in either an oven using hot air or by passing the film over a series of heated rolls, as is the case in Figure 2.13. Great care must be taken to insure a constant temperature across and along the film.

Film Stretching

With minor exceptions, all flat film stretching is done sequentially: that is, the film is stretched first in the MD and then in the TD. The main reason for this sequence is that MD stretching is accomplished by drawing the film between rolls moving at different speeds. If this were done after the film is stretched TD to its full width, the MD stretching rolls would have to be much wider (4X to 8X) and more massive. While such machines exist, they are much more expensive and designed for specialty films. There are machines available that stretch a flat film simultaneously in both directions, but being mechanically very complicated, they are not only more expensive but require much more precise and costly maintenance as well. Hence, they too tend to be used only for specialty films.

As shown in Figure 2.13, MD stretching is accomplished by drawing the reheated film between two sets of heated rolls with the downstream set moving at a higher speed. Since the force required to draw the film is high the film must be gripped well by having the film contact as much roll surface as possbile or by using nip rolls. Also the surface must be designed to balance between adequate gripping force on the film, nonmarring of the film surface, and, for rubber covered rolls, reasonable life before replacement.

The machine used to stretch plastic films in the TD is called a tenter frame and consists of two continuous chains like two race tracks mounted side by side. Clamps, like the one shown in Figure 2.14, are mounted on top of each link and these grip each edge of the film. The film, then, is carried along as the chain is driven forward. The two chains gradually move apart, and as they do they draw the film in the TD between them. The chain rides on a track on well-lubricated sliding surfaces or ball bearings at speeds up to 1000 feet per minute. At the end of the stretching section, the chain releases the film, turns around a wheel and is returned to the beginning of the TD stretching section where it reverses direction again. This constantly changing momentum puts heavy demands on the wear resistance of the metal contact surfaces.

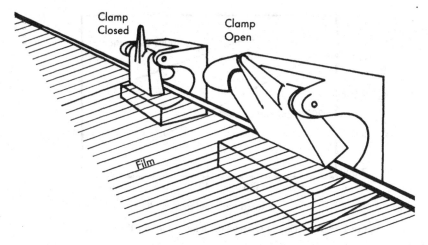

Figure 2.14 Tenter Frame Clamp.

The clamps that ride on the chain must be able to grip the film securely as the film enters the TD stretching section, hold it against the stretching force applied to the film and then release the film as it leaves the TD stretching section. This gripping and releasing is accomplished by having the jaw pivoted so that its weight causes it to be normally in the closed position. To open the clamp, a fixed arm strikes a finger on top of the clamp at the appropriate locations at the beginning and end of the stretching section. The gripping force of the jaw is just enough to grip the film and yet not enough to damage the edge, which could lead to a tear that would propagate across the sheet. This difficult balancing act is usually aided by creating a heavy edge or "bead" on the film as it is formed at the film die. Before the film reaches the wind up, the bead must be slit off to avoid built-up edges on the winding roll and film distortions as described earlier for the cast film process.

The tenter frame is enclosed in an oven that is usually heated with hot air. The temperature of TD stretching is generally higher than that of MD stretching, since during the MD stretching some crystallization has been induced that increases the resistance to further stretching. In fact, the oven is typically separated into zones where different temperatures can be maintained for optimum stretching. Within each zone, temperature uniformity must be maintained so that the sheet stretches uniformly.

The film responds to the applied forces as follows. As temperature is increased the degree of molecular orientation achieved for a given amount of stretching decreases. This is because molecular relaxation begins to compete more favorably with the crystallization. This effect is shown in Figure 2.15 where two levels of stretch (200% and 400% elongation) of poly-

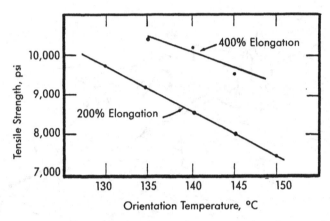

Figure 2.15 The Effect of Orientation Temperature for Polystyrene.[8]

styrene were applied over a range of temperatures. Tensile strength is a measure of the level of molecular orientation achieved. Note that at a given temperature tensile strength increases sharply at the higher stretch level and falls off at the same stretch level as temperature increases. However, there is a limit to the stretching temperature since as the temperature decreases the stretching force increases to the point where machine capability or film strength is exceeded.

Given the sensitivity of orientation to temperature, nonuniformities in temperature in the sheet cause variations in orientation and in physical properties. There are counterbalancing effects. When a crystallizing film is stretched, there is an increasing resistance to stretch since orientation accelerates the crystallization. This effect limits the degree to which a particular polymer system can be stretched, and it also tends to reduce variations in orientation due to localized differences in temperature. As warmer areas stretch more readily, they also crystallize faster shifting the stretching force to the cooler areas. Similarly, an area that is already thinner stretches more at first but because of accelerated crystallization, the stretching is shifted to thicker areas. In addition, TD stretching can be done with a decreasing temperature downstream. Since thin sections cool faster, stretching is again shifted to thicker areas.

Despite these self-correcting effects, orientation causes a magnification of the thickness variations in the cast sheet by as much as 3 to 5 times. Therefore great care is taken to achieve the best possible uniformity in the cast sheet. Temperature uniformity throughout and stable mechanical operation are also essential. Given all the process variables that can affect thickness variation, several measurements must be combined to define

[8]Benning, p. 20.

corrective action. Normally the thickness is measured continuously in the MD and TD for both the cast film and final oriented film. Computer analysis is widely used, and closed loops to automatically adjusting film dies are increasingly being adopted. Finally, thickness variations are usually randomized in the final roll by oscillating the wind up stand.

Film Heat Setting

This step is accomplished in the last section of the oven. The temperature of this section is elevated and often the tenter chain rails are brought a little closer together. This permits a slight relaxation in the film. Also the higher temperature increases the perfection of crystallization.

TUBULAR ORIENTATION PROCESS

The two major drawbacks of the flat, oriented film process are its mechanical complexity and the imbalance of properties because the stretching is sequential. In a tubular process (Figure 2.16) the stretching is done by inflation with air pressure which avoids much mechanical complexity. In many respects the tubular orientation process is similar to the blown film process. One major difference is that the preferred direction for tubular orientation is downward. A second difference is that the extruded tube is quenched and then reheated before inflation in analogy with the steps described for the flat, oriented film process above.

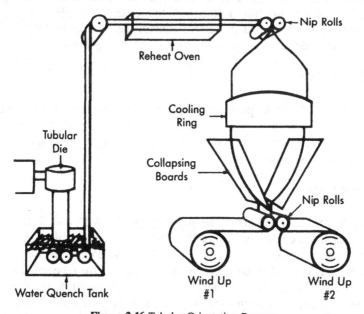

Figure 2.16 Tubular Orientation Process.

Melt Film Formation

The formation of a tube from the extruded melt is accomplished just as in the blown film process described earlier. The circular die is sometimes more complicated, however, when the core must make room for the cooling water and the other components of a mandrel quenching system that are fed through the die. This increases dramatically the complexity of the die (for example, increasing the number of internal seals) especially if the die is rotated to randomize thickness variations in the emerging melt.

Melt Film Quenching

As in flat film processing, quenching must occur as rapidly as possible. In the system shown in Figure 2.16 this is accomplished by a water bath. In other systems a sheath of cooling water contacts the outside of the melt tube. This water is removed downstream using squeegees. The inside surface can be cooled as well by a layer of water injected between an inner mandrel and melt tube. Another approach is to use a quench mandrel, which is a cooled metal cylinder mounted on the die so that the melt tube as it emerges from the die and contracts will collapse onto the mandrel, establishing intimate contact. Regardless of approach, the quenching step must be accomplished as uniformly as possible to minimize thickness variations in the subsequent step, in which orientation is accomplished by MD stretching and TD inflation. It is necessary to prevent both the drawing force and the air pressure from getting back to the weak melt film. The nip rolls in the quench tank and those that contain the bubble as shown in Figure 2.16 provide the necessary isolation of stretching forces.

Film Reheating

This is accomplished by passing the tube through an oven with radiant heaters. Uniform heating is essential to achieve uniformity of stretching. Heater output and spacing must be identical and the alignment of the tube with the oven must be precise.

Film Stretching

Although stretching occurs simultaneously in the MD and TD, the forces for each are controlled separately. The MD force is applied by the differential speed between the two sets of nip rolls that contain the bubble. The TD force is applied by air pressure introduced into the tube in a variety of ways as discussed for the blown film process. To minimize thickness variations, cooling air is often used on the outside of the inflating tube to shift the stretching to the thicker sections of the tube.

Control of the final tube diameter is usually accomplished by using a metal cooling ring. This fixes the maximum film width for the sequence. This final film width is not easy to change in the tubular process since it requires changing both the die and quench mandrel as well as the cooling ring if the TD stretch ratio is to be kept the same. In the flat film process, width is more easily altered by adjusting the cast film width with end blocks on the die and moving the tenter frame rails in or out.

After the tube passes through the cooling ring, it is collapsed, flattened, and slit into two sheets exactly as in the blown film process. Winding onto rolls is also the same and all of the principles, described in the section on the cast film sequence, to achieve uniform winding are applied here as well.

Film Heat Setting

Heat setting, necessary to decrease residual shrinkage in the oriented film, is the Achilles heel of the tubular process. Heat setting must be carried out at a higher temperature than the stretching step, and a higher pressure is required to sustain the fully inflated bubble. The only feasible way to accomplish this on the inflated bubble is by means of the "double bubble process". Here, the originally stretched tube is collapsed and flattened by a pair of nip rolls. These nip rolls isolate the higher pressure of the heat setting step from the stretching step. The tube is then reinflated and passed through a second oven where heat setting occurs prior to final collapsing, flattening, and slitting. This approach greatly complicates the operation and has not proved as effective as desired. Consequently, heat setting is generally achieved as an added step after the tube has been slit, either just before the windup or off-line using a separate machine. In either case, reheating of the film is accomplished by passing the film over heated rolls. Shrinkage occurs in both the MD and TD. MD shrinking is easily accommodated by reducing the speed of the downstream rolls. TD shrinkage, however, leads to slippage of the film on the surface of the rolls which can cause scratching.

In the sections that follow, the two orientation processes will be compared as to film properties and range of application.

FLAT AND TUBULAR ORIENTATION COMPARISON

Process Dependent Properties

The most significant difference between the two processes is that in tubular orientation the MD and TD stretching are simultaneous so that properties that depend on orientation are closer to equal in both directions. In sequential stretching, the orientation induced by the last step dominates, leading to strength properties that are higher in the TD except for tear strength which is always weaker in the direction of highest orientation. It is as if some of the original MD orientation is pulled out as the sheet is subsequently stretched in the TD. On the other hand, flat film stretching with its mechanical constraints on the film is more uniform. Therefore, thickness variation and distortions in film flatness tend to be greater in the tubular process. The difference can be minimized by randomizing irregularities through rotation of any or all elements of the equipment, but especially the windup and the collapsing sections.

Another difference is that the bead, required for the flat film sequence, is eliminated in the tubular process. Furthermore, edge effects that distort properties close to the tenter frame rails are also absent in the tubular process. Finally, dimensional stability is more readily achieved in the flat film sequence without increased wrinkling and surface scratching.

This comparison is summarized in Table 2.2.

Range of Application

For both processes, only polymers that can be cast into their amorphous state can be successfully oriented. Polymers of interest to packaging fall into three categories. The first category is highly crystalline polymers that are readily heat-stabilized to form dimensionally stable films. These include PET, PP, nylon, and EVOH. The second category is polymers with reduced crystallinity that yield films with high levels of residual shrinkage. These polymer systems are especially suited to shrink applications and include ethylene/propylene copolymers, polyethylene blends, LLDPE, and plasticized PVC. The third category is amorphous polymers which, since they cannot be heat-stabilized, shrink back to their original dimensions when heated.

For packaging applications using heat-stabilized films, a lack of balance in the MD and TD properties of oriented films is not a problem. Therefore, given the difficulties of the tubular process in achieving good thickness uniformity and sheet flatness and especially the problems with heat

Table 2.2. Comparison of flat and tubular orientation.

Property	Flat	Tubular
MD/TD Balance	−	+
Thickness Uniformity	+	−
Film Flatness	+	−
Randomization		+
Bead and Edge Effects	−	+
Dimensional Stability	+	−

setting, the flat film orientation sequence is generally preferred. In practice, only OPP films are made by the tubular process. While attempts have been made over the years to make PET films this way, no significant production exists today. However, the development of machinery for both approaches has advanced over the past decade so that complete film lines can be purchased for either process. The primary consideration for choosing one over the other is economics, which will be covered later.

For shrink films, most applications require balanced MD and TD shrinkage. Since the flat film sequence produces oriented films with residual shrinkage primarily in the direction of the last stretching step, the tubular process is preferred. The difficulties of heat setting tubular films is not a problem since this step is eliminated. The film undergoes heat setting later when it is heat shrunk around the package contents (see Chapter 5). Therefore, almost all commercial production of shrink films is by the tubular process.

FLAT AND TUBULAR ORIENTED FILMS— TYPICAL SYSTEMS

In this section, only those film properties that are especially affected by the orientation will be discussed. For a more complete summary of properties the reader is referred to Table A.1.

Polyethylene Terephthlate (PET)

This polymer was the first to be used to manufacture a biaxially oriented film on a large commercial scale. It readily quenches to an amorphous state. Maximum stretch ratios are about 3.5X MD and 5X TD. Orientation takes place above 80°C and heat setting at about 200°C. Modern, wide lines producing film up to 240 inches in width and at speeds of 1000 feet per minute using the flat film orientation process are now common, and recently a Japanese line 320 inches wide has been produced. Most packaging applications utilize 0.48 mil film with a range from 0.35 to 0.92 mils.

Orientation increases the tensile strength of PET film from 6,000 to 25,000 psi and reduces the elongation to break from 500 to 100%. The tensile modulus is quite high at 700,000 psi. On the other hand, the force to propagate a tear is low: 15 grams. A major advantage of oriented PET film is its retention of tensile properties at temperatures up to 150°C. At 150°C shrinkage is only about 2% in 30 minutes. Flex resistance (as reflected in the pin hole flex test) is not as good as it is for some other oriented films (especially nylon film) due to the highly developed lamellar structure. Orientation significantly increases oxygen barrier for PET film. In one example a stretch ratio of 4X doubled the barrier. Even so, while the oriented film is a reasonably good oxygen barrier, its moisture barrier is intermediate among oriented films.

Polypropylene (OPP)

OPP is by far the most widely used oriented film for packaging applications. Stretch ratios for OPP can be higher than those for PET, with a maximum of about 6X by 6X. Film is produced from homopolymer and from copolymers with 0.1 to 1.5% ethylene using both the flat and tubular orientation processes. Extrusion takes place at 200°C with orientation at 150 to 160°C. For the flat film process, thickness ranges, film speeds and widths are all in the same range as given above for PET. For the tubular process, maximum tube circumference is 240 inches.

The flat film sequence produces films that are more uniform in thickness and in sheet flatness. There is little difference in the physical properties between the two approaches. Tensile properties are more balanced for tubular film. For example, for one set of films the tensile strength was 26,000 psi for both MD and TD for the tubular process film but 17,5000 MD and 42,000 psi TD for the flat process film. Dimensional stability is poorer for tubular films which, in the same example, shrank at 124°C in 60 minutes, 8% MD and 15% TD as compared to a balanced 3.5% for flat film.[9]

Nylon

Oriented films of both 6,6 and 6 nylon have been produced and offered commercially, and their properties are quite similar. Maximum stretch ratios are typically low at 3X MD and 4X TD. Given the lower volume of sales, commercial lines tend to be smaller than for PET or OPP. Film thicknesses are in a range similar to PET films. Both flat and tubular orientation processes are used. The usual effects of orientation are seen: tensile strength increases from 11,000 to 30,000 psi and modulus from 100,000 to 300,000 psi; tear strength is 15 versus 30 grams for the unoriented film; and oxygen barrier is improved by about 25%.

Polystyrene (PS)

Both homopolymers and copolymers are oriented by the flat and tubular processes. Unoriented films are brittle but oriented films have a high tensile strength, an improved impact strength, and a high modulus. Produced

[9]Park, W. R. R. 1969. *Plastics Film Technology*. New York, NY: Van Nostrand Reinhold Company, p. 29.

on a small scale, a typical flat film process casts film 15 inches wide and 90 mils thick from an extrusion at 220°C. After quenching at a relatively high temperature (105 to 130°C), the film is biaxially stretched 3X by 3X yielding a final film 45 inches wide and 10 mil thick at a speed of 30 feet per minute. Tensile strength is increased by orientation from 8,000 to 17,900 psi. Shrinkage at 150°C is 5% MD and −2.9% TD.[10]

Polyvinylidene Chloride (PVDC)

A typical PVDC coplymer used is a 85/15 vinylidene chloride/vinyl chloride composition containing 7% di(alpha-phenyl ethyl)ether. Both flat and tubular processes are used and some film is made using solvent casting for the flat film process. Normally extrusion takes place at 170°C. For the tubular process this is followed by water quenching, with mineral oil on the inside of the tube to minimize blocking. This is a small scale operation with bubble diameter at 12 inches and stretching at room temperature to 3–4X MD and 2.5X TD. For 2 mil film, throughput is at 75 pph. The film is heat shrinkable above 80°C. The double bubble process is used to produce a heat stabilized film.[11]

Shrink Films

This is a class of oriented films that exhibits high shrinkage (25% or more) at relatively low temperatures (less than 100°C). Generally, polymer systems with reduced crystallinity are employed. Another aspect is that in the sequential stretching of the flat film process, the absence of adequate crystallinity to stabilize the orientation produced in the first step causes the second direction of stretch (typically TD) to dominate even more, so that for most shrink film polymers it is very difficult to produce a film with balanced shrink properties. Therefore the tubular process is the one most often used for shrink films.

For good shrink properties, some polymer systems must be modified. For polyethylene various blends, copolymers such as LLDPE, and/or irradiation are used. For polypropylene, copolymers with ethylene must be used. PVC and its copolymers are plasticized.

The tubular orientation process is the same for shrink films as that described earlier except that the heat stabilization step is eliminated. For polyolefins, processing conditions, tube sizes and throughputs are similar to those described for OPP. In the case of polyethylene, irradiation for cross-linking the cast tube can be added to the sequence prior to stretch-

[10]Ibid.
[11]Ibid. p. 35.

ing. Normally, unmodified LDPE cannot be oriented since the stretching temperature is too close to the melting point; however, cross-linking leads to a temperature range that is adequate. Another gain from cross-linking is a broader range of shrink temperatures.

For PVC, the flat sheet orientation process is used as well as the tubular process. In one variation, the sheet is cast from a solution and then stretched by a tenter frame in a separate operation.

The properties of shrink films depend largely on the polymer system used. Tensile strength, for example, ranges from 9,000 psi for LDPE to 26,000 for polypropylene. Also, the temperature range over which significant shrinkage occurs varies from 65–120°C for polyethylene to 120–165°C for polypropylene.

ECONOMICS

In this section, the economics for the film processes described above will be discussed, primarily in semiquantitative and relative terms. This approach has been chosen because precise absolute numbers are soon out of date and because a more qualitative treatment is more instructive for the general reader.

Investment

Although the investment for bare equipment only will be used in these comparisons, it must be remembered that the complete investment for an operation also includes

- cost of installation (which can be as much as the bare equipment cost)
- cost of adding services such as electricity, compressed air, and air conditioning
- cost of auxiliaries such as slitters, materials handling, and packaging lines
- cost of buildings

Where all these additional items are needed, the total investment can be 3 to 4 times the bare equipment cost.

For cast film lines, the investment is about $800 to $1,000 per pound of film produced per hour, with the largest lines having the lowest unit investment. Large lines costing about $1 million produce about 1,200 pounds of film per hour whereas a line producing at about 1/3 that rate costs about half as much. The output of blown film lines is about 10 to 20% lower for the same investment, but this output and investment gap will probably be

closed in the future. Film yields are in the 90 to 95% range for both processes.

The investment for oriented film equipment is higher for two reasons:
additional equipment and lower yields. For a tubular orientation line, the
bare equipment cost is 2 to 3 times higher while the yield is in the 60 to
70% range. These combine to make the unit investment 3 to 6 times that
for a simple blown film line. In the flat film process the tenter frame is not
only large and mechanically complex but also requires the handling of
much greater volumes of hot air. As a result, the investment is higher for
oriented flat film lines versus oriented tubular film lines. However, the
tubular process is limited by heat transfer at a lower film speed than is the
flat process. The result of all these factors is that the flat film process represents a somewhat lower unit investment for large machines while the
tubular process is less expensive for smaller machines.

Manufacturing Cost

The two basic elements of manufacturing cost are the cost of materials
and conversion cost. The latter includes the labor, energy, and the over-
head directly associated with the film operation. Depreciation and the
costs of auxilary operations such as slitting, packaging and shipping are
excluded. Including those costs would raise the total to 2 to 3 times that of
the conversion cost shown here.

The conversion cost for cast and blown films is very similar at about
$0.06 per pound ($\pm$ 10%). Since material costs vary from about $.45/lb for
LDPE to $0.70/lb for nylon, conversion cost is about 10 to 15% of the
material cost. Given the lower yield for oriented films and the increased
labor required by the more complex operation, conversion cost is about
double that for unoriented cast and blown film. The lower yield for
oriented films is not a significant factor in material costs since essentially
all of the film waste is recovered and recycled.

Summary

Cast and blown film operations are generally relatively small and often
serve regional customers. Oriented film processes, on the other hand, with
their higher investment and cost and technical complexity are usually operated on a larger scale by companies with a depth of experience in making
plastic films. However, now with the ready commercial availability of
small tubular units it is becoming easier for smaller manufacturers to get
into oriented films especially for less demanding markets such as some
shrink film applications.

BIBLIOGRAPHY

Bakker, M., ed. 1986. *The Wiley Encyclopedia of Packaging Technology*. New York, NY: John Wiley and Sons, Inc.

Briston, J. 1983. *Plastics Films. Second Edition*. Essex, England: Longman Group Limited.

Griff, A. L. 1962. *Plastics Extrusion Technology. Second Edition*. New York: NY: Reinhold Publishing Corp.

McKelvey, J. M. 1962. *Polymer Processing*. New York, NY: John Wiley and Sons, Inc.

Park, W. R. R. 1969. *Plastics Film Technology*. New York, NY: Van Nostrand Reinhold Company.

Tobolsky, A. V. and H. F. Mark., ed. 1971. *Polymer Science and Materials*. New York, NY: John Wiley and Sons, Inc.

Vinogradov, G. V. and A. Ya. Malkin. 1980. *Rheology of Polymers*. Moscow, USSR: Mir Publishers.

CHAPTER 3

Processes for Modifying Plastic Films

INTRODUCTION

In general, plastic films as produced by the various processes described in Chapter 2 do not possess all of the properties required for demanding applications such as food packaging. Exceptions include most applications for shrink films and for blister and skin packaging as well as pouches from cast or blown films for non-food applications. However, many of these less demanding applications require surface printing and plastic films, generally, have relatively inert surfaces that do not provide sufficient adhesion to printing inks. Therefore, some kind of surface treatment is required. For food applications, gas barrier greater than most plastic films can offer is often required, so a layer of high barrier polymer is added. For the oriented films in particular, the ability to form a heat seal must be added by using a layer with a lower melting point.

This additional processing of the film is typically done using a separate machine and often by a manufacturer (a converter) different from the resin or film maker. However, when the added complexity can be tolerated, it is sometimes more economical to insert this added step as part of the film formation sequence. When a separate machine is used, the film must be unwound from the finished roll, fed through the sequence of processing steps and then wound again into a roll. All the procedures required to achieve a uniform, wrinkle-free roll apply here; the focus in this chapter will only be on the added steps.

SURFACE ADHESION

Most of the film processing steps that will be described in this chapter add another material or film to the original film in order to improve or bring new properties to the final structure. The ability of the plastic film to adhere strongly to inks, coatings, etc. is critical. In order to understand how this adhesion is achieved, measured, and controlled, it is necessary to develop a fundamental picture of this property.

Adhesion in Plastics

The first principle of adhesion is that there is always a force of attraction between two contacting surfaces. This force arises from the electronic configurations within the molecules, which at some part of a site will create a predominantly electronegative environment with a corresponding electropositive environment at another part. These localized areas of separated electropositivity and negativity are called dipoles. Like magnets the positive end of one dipole will attract the negative end of another one. For all molecules this effect occurs for at least brief instants in time since among all the possible electron configurations that occur over time there are some that will create transient dipoles. While this may seem to be a small effect, the number of molecules in a gram of a material is more than 10^{20} and so the cumulative effect is sufficiently large to account for the force of adhesion for all hydrocarbon polymers such as polyethylene and polypropylene. When other elements are part of the polymer molecule, such as oxygen and nitrogen atoms, the chemical bonds involved are permanent dipoles. These contribute considerable additional attractive force for the dipoles in other molecules. Further, the electrical field associated with these dipoles can induce transient dipolar properties in other molecules, which in turn leads to an attractive force. Finally, when certain forms of oxygen and nitrogen bonds are in the vicinity of certain hydrogen atoms, an especially strong attraction occurs called hydrogen bonding. While the strength of this bond is only 5 to 10% of a carbon-hydrogen bond it is still the strongest type of adhesive force between polymers short of a chemical bond. Examples of groups that readily form hydrogen bonds are carbon-oxygen (carbonyl), carbon-nitrogen (cyanide) and carbon-chlorine. Groups with hydrogen atoms that readily participate in these hydrogen bonds are, for example, oxygen-hydrogen (alcohol) and nitrogen-hydrogen (amine). As will be seen in later sections most modifications of a plastic film to increase adhesion involve the introduction of these dipolar or hydrogen bonding sites onto the surface either by some treatment or by coating.

Given that attractive forces of some kind exist between all polymer molecules, the next consideration is how to maximize these forces. The force of attraction between two flat surfaces falls off as the third power of the distance between them. Thus, even though the total force of attraction across an interface is the composite of the attractions between all of the molecules of one body and those of the other, only the interactions very near the interface will contribute significantly to the interfacial adhesion. From a practical point of view this means that very intimate contact, usually liquid contact, is essential. In fact, the ideal interface for maximum bonding is one that allows diffusion across the interface and polymer molecules from one body intermingling with those of the other. It also follows that any interfering layers such as air, oil, or dirt must be eliminated.

The Measurement of Adhesion in Plastics

To determine how effective any combination of the above factors is in achieving adhesion between two surfaces, the force of adhesion must be measured. For two solid blocks, this is simply a matter of determining the force required to pull the blocks apart either directly (forces applied perpendicular to the plane of interface) or in shear (forces applied parallel to the interface). The measurement is more complex for plastic films since they are thin and flexible. To measure, for instance, the adhesion of one film heat sealed to another, a block could be adhered to each of the opposite surfaces of this structure. If the adhesion of the film surfaces to the blocks greatly exceeds the adhesion between the films, then the force measured is approximately the force of adhesion between the films. To measure exactly the force of attraction a correction for that part of the force that is deforming the various plastic layers needs to be made. With precise control of conditions and separate measurements of the deformation, it is possible to distinguish between these. However, in practice, only the total force to separate is measured since in fact, for most packaging applications, this is what is important. Thus to maximize a bond strength, increasing deformability is often employed in the formulation of adhesives for plastic films.

In addition to the ability to absorb energy by deformation, plastics have another advantage over many other solid materials that helps achieve high levels of adhesion. Since most plastics can be melted, it is possible to bring together two liquid surfaces (as in heat sealing) for maximum contact and to eliminate the air layer. Other surface contaminants can be absorbed into the liquid if they are compatible with it. Alternatively, as in the case of melt coating, a liquid can be brought into contact with a solid surface. Finally, for high melting polymers, it is often possible to dissolve the coating polymer in a solvent or disperse it in water.

The measurement of adhesion between two plastic films using two blocks is not practical in the commercial environment and, in addition, the measurement does not accurately simulate the forces at play in a film application. In a pouch, for example, it is the force required to peel the layers of the heat seal apart that is of interest. The measurement of adhesion is made, therefore, by pulling apart the two films as shown in Figure 3.1 in two typical configurations. This simple procedure involves a complex

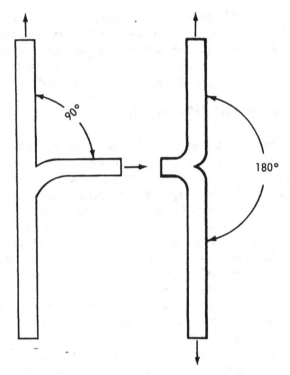

Figure 3.1 Peel Test for Adhesion: Typical Configurations.

combination of forces since the films are being deformed by bending. Since the peel force changes with the peel angle, it is important to insure that the angle between the heat seal and the separated films is reproduced. Another important aspect in measuring peel force is that it is not constant even though the rate of peel is carefully controlled. Instead, it fluctuates sharply in a stick-slip fashion because of the randomness of imperfections at the interface. The peel force is generally stated as the average of these often widely varying values.

So far, only the example of one film heat sealed to another has been discussed. Of interest as well is the adhesion between a thin coating (for example, printing ink or a barrier coating) and a film substrate. Here, a second film is adhered to the coated surface and then the two films are peeled apart as described above. Often this second film is pressure sensitive tape and, for many applications, the adhesion of the coating to the substrate is considered to be adequate if it exceeds the adhesion of the coating to the tape.

Whether peeling apart two heat sealed films or a pressure sensitive tape from a coated film, it has been assumed so far that the peeling is occurring by breakdown of the film/film or coating/film interface. In fact, however, there can also be cohesive failure within the films or coating layer. Thus, it is essential to determine the identity of the two surfaces that have pulled apart. In Figure 3.2 the composition of these surfaces could be three kinds assuming it is not the tape that is peeling off.

First Surface	Second Surface
1. Substrate film	Coating
2. Coating	Coating
3. Substrate	Substrate

For (1) there is a clean break between the coating and the substrate film and the weakest link is the adhesion between the two. For (2) there is a cohesive failure in the coating (that is, the coating deforms so easily to the applied force that its cohesive strength is exceeded and it ruptures). For (3) there is a cohesive failure in the substrate film. This sometimes happens in oriented films since the strength in the thickness direction of the film is greatly diminished. These distinctions point to the area where improvement must be made for increased adhesion.

FILM TREATMENT FOR ADHESION

If the failure in adhesion for a printing ink or coating to a film is at the surface of the film, then often it is found advantageous to treat the film surface. This treatment could be simply a roughening of the surface, thereby

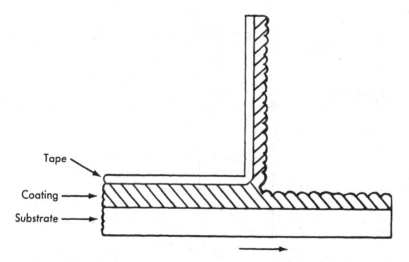

Figure 3.2 The Measurement of Adhesion for a Coated Film.

increasing the surface area of contact between the initially liquid ink and the film, but more often a change in the structure of the surface is required. One change is in the crystallinity. Highly crystalline surfaces tend to be rather impenetrable and any polar groups along the polymer chain are less available at the surface to promote bonding. However, for many polymers, such as polyolefins, there is insufficient inherent polarity, and a change in the chemical structure of the surface is required.

Corona Treatment

The most commonly used surface treatment for plastic films is to create a corona discharge in the vicinity of the film surface. This is accomplished by imposing a large voltage drop between an electrode and an insulated roll over which the film is moved. A diagram of a typical arrangement is shown in Figure 3.3. The gap is small, usually about 0.5 to 1.0 inches, and must be uniform over the width of the film. With the right power supply, a stable, uniform corona discharge can be established that consists of a plasma of ionized gas and other reactive species. The primary effect is the interaction of free radicals in the plasma with the film surface leading to reactions such as oxidation and polymer chain scission.

The level of treatment depends on the concentrations of the active species in contact with the surface, which depend in turn on such variables as

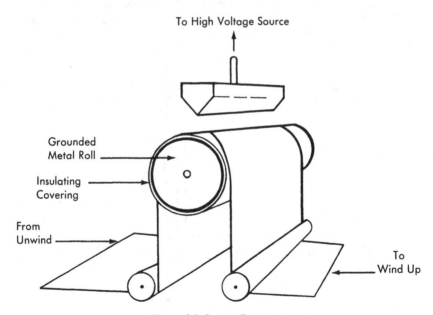

Figure 3.3 Corona Treatment.

duration of treatment and voltage. For example, as time of treatment was increased from 1 to 10 seconds, the bond strength of a polyamide adhesive on polyethylene film was increased from 3 to 16 kg/cm^2.[1] In another example a voltage increase from 3150 to 8650 volts increased the rate of production of active species about 15 fold.[2]

Normally, the corona is created in an atmosphere of air so that the process is easily installed as a step in the film formation sequence. However, the process generates toxic ozone so the atmosphere in the treating step must be contained and exhausted from the operating area. At high film speeds, a layer of air that contains ozone adheres to the film as it leaves the treating enclosure. This layer must be removed from the film by close-fitting baffles or a vacuum system.

In some cases, creating the corona in special atmospheres such as solvent gases and rare gases leads to even higher chemical activity at the surface and higher levels of adhesion. This requires special chambers and is almost always carried out off-line.

Flame Treatment

For this treatment, a flame is impinged directly onto the film surface as the film is moved across a cooling roll. Either an oxidizing or reducing flame can be produced, depending on the concentration of air or oxygen in the gas feed. These flames contain excited species that chemically react with the polymer surface, producing free radicals that in turn primarily undergo oxidation. Due to the safety hazards and the increased complexity of controlling the flame-treating process, it is not widely used for packaging films.

Priming or Sub-Coating

For priming or sub-coating, a very thin coating of a polar substance is applied to the film surface, often from a water dispersion. General types of polymers used for this purpose are polyalkyleneimines, polyurethanes, and polyesters. Organic titanates and colloidal silica are also used. The priming step is often included in either the film formation sequence or the coating or printing steps. If the sub-coatings are tacky, making it difficult to wind and unwind film that has been primed, carrying out the priming as part of printing or coating allows the sub-coat to be completely covered by the later coatings prior to winding. Where solvents are used for the

[1]Wu, S. 1982. *Polymer Interface and Adhesion*. New York, NY: Marcel Dekker Inc., p. 317.
[2]McKelvey, J. M. 1962. *Polymer Processing*. New York, NY: John Wiley and Sons, Inc., p. 167.

Table 3.1. The effect of corona treatment on the
adhesion of polyethylene.[3]

	Adhesion (g/in)	
Substrate	Untreated	Treated
Polypropylene	0	>900
Polystyrene	0	600–800

priming solution, they must be recovered or incinerated in an approved manner.

The application of these coatings must be accomplished in a way that insures complete coverage of the surface and reasonable control of thickness uniformity. Typically, printing techniques are used such as using gravure rolls and offset rolls. These are described later in the section on printing.

Process Dependent Properties

The effect of corona treatment on the adhesion of a coating of 1 mil polyethylene applied at 250 fpm on two substrates can be seen in Table 3.1. Even where there is some adhesion to an untreated film, treatment often increases the level. An example of this is shown in Table 3.2 where a 0.12 mil coating of vinyl chloride/acrylonitrile copolymer (80/20) was applied to OPP. The effect of different 1 mil thick primers on the adhesion of a 1 mil polyethylene coating to PET film is seen in Table 3.3.

Since only the surface of the film is affected by these treatments, no change in physical properties is involved. In cases of extreme levels of treatment, some discoloration can occur.

Range of Application

As implied by the examples above, almost all plastic films exhibit some level of improvement in adhesion, especially to polar printing inks and

Table 3.2. Corona treatment of OPP.[4]

Film	Heat Seal Strength (g/in)
Untreated	400
Corona Treated	680

[3]Noll, P. B. and J. E. McAllister. 1963. *Film and Foil Converter* (October), p. 46.
[4]British Patent 920,078, March 6, 1963.

Table 3.3. The effect of different primers on the adhesion of
polyethylene to PET.[5]

Primer	Adhesion (g/in)
None	0
Polyethyleneimine	844
Polydibutyl titinate	504
Colloidal Silica	0

coatings, as a result of corona treatment. Since corona is the easiest and most inexpensive treatment to apply, it is normally attempted first. When higher adhesion levels are required or when the adhesion must resist fairly severe conditions such as high temperature or high humidity or boiling water, then more chemically active treatments such as corona in special gases or flame treatment must be used. Priming is normally the most reliable solution to a difficult adhesion problem since the coating can be formulated to provide exactly the right chemistry for adhesion to both the substrate film and to subsequently added coatings.

PRINTING

Many film packages found on retail shelves are printed. In the simplest cases, black lettering is used for identification and instructions. In the most complex cases, the entire package surface is covered with up to six colors picturing the contents and/or giving an advertising message. Since printing technology was developed initially for paper, it has taken substantial effort and time to reach the level of precision and complexity that is routine today for plastic films. The primary properties of plastic films that had to be taken into account were their extensibility under tension, the inertness of their surface, and their tendency to generate static electricity when processed at high speeds.

Surface inertness was overcome partly by surface treatment. Another assist came from the reformulation of the printing inks using polymers that were more compatible with the plastic film surfaces. As a result, a wide range of ink systems is now available from different suppliers. Usually the optimum choice is determined empirically from the results of wetting and ink adhesion tests.

The extensibility and static electricity problems were overcome by redesigning the printing presses. All of the principles described in Chapter 2 for moving films at high speeds and controlling tension apply to printing presses as well: careful control of drying temperatures, tension isolation

[5]DeHoff, G. R. and T. F. McLaughlin, Jr. 1963. *Modern Plastics* (November), p. 107.

and minimization, and unwinding and winding rolls with programmed tension. The generation of static electricity in plastic films is a problem because printing inks are often formulated with flammable solvents. Given these safety concerns and the desire to eliminate solvents from the environment, the trend is to replace solvent inks with systems based on aqueous dispersions. While dispersion inks do not yet achieve the standard of quality set by solvent inks, much progress has been made in recent years.

While simple printing jobs are sometimes done as an added step in the film formation sequence, a separate machine is usually employed. The heart of the machine is the printing station, and several different techniques are available. Flexographic printing is the most widely used technique for packaging films. A diagram of this process can be seen in Figure 3.4. A roll contains a raised image on a flexible rubber plate. This image is usually produced on the flexible plate by a photochemical process that transfers the image from a negative. Thus the cost of such plates is relatively low. In the printing process the impression roll receives ink transferred onto it via a series of rolls from a reservoir. The plate is brought into contact with the moving film, which is pressed onto it by backup impression cylinder.

The next most important process is rotogravure printing. Here the image, which is engraved (rather than raised) into a steel roll, consists of a series of dots that are filled with ink. When the roll contacts the film, the

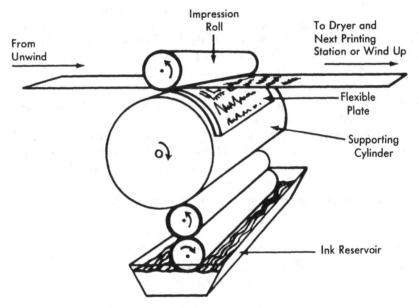

Figure 3.4 Flexographic Printing.

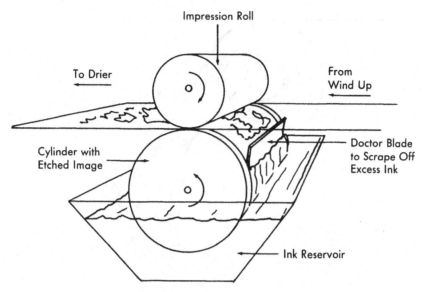

Figure 3.5 Rotogravure Printing.

ink is transferred to create the image. Since the depth of the dots can be altered over a wide range, more precise control of tone is achieved. Ink is applied directly to the printing cylinder as shown in Figure 3.5. The doctor blade is used to scrape off the excess ink.

Using either of the printing rolls described above, a single color ink can be applied repeatedly along the length of a moving film. If additional colors are required, additional printing stations are employed. Up to six colors are applied by the most sophisticated presses. Additional colors are created by appropriate combinations of two colors (for example, green from an overlay of blue on yellow). Precise registry of these multiple impressions is essential to achieve the high quality of the final image that appears on the many packages exhibited on the store shelves today.

Other steps in the printing process involve the unwinding of rolls of film, the control of tension throughout the path of the moving film, the drying of the film, and the final winding on the finished roll. Frequently, the number of impressions desired for a particular job exceeds the film area available on a single roll of film. In this case, a step is added to allow the smooth transition from the expiring roll to a new, waiting full roll. This eliminates the loss in time and startup waste that accompanies interrupting the printing operation for a roll change. A turret winder, like that shown in Figure 3.6, allows easy transfer of the unwinding film from an expiring roll to a new roll. To prepare for the transfer, the turret is rotated so that the surface of the new roll just contacts the moving film coming from the

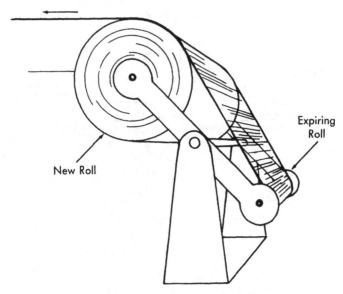

New Roll

Expiring
Roll

Figure 3.6 Turret Winder in Position for a Roll to Roll Transfer.

expiring roll. The speed of rotation of the new roll is matched to that of the film and a sticky substance is then placed on the surface of the new roll so that it is grabbed by the moving film and pulled through the press. This transition section, which consists of two layers stuck together for the length of the overlap between the old and new rolls, must be removed from the finished roll at the end of the press. Usually during this transition, the printing rolls are pulled back away from the film so that the printing stations are not disturbed.

When printing is applied to a monolayer film, such as a polyethylene film to be used by itself to form a pouch, it is the last step in the sequence of film processing. In contrast, for more complex coated or laminated film structures, printing is usually applied early in the sequence so that the printing is buried in the structure by the subsequently added layers. In this way the printing is protected from abrasion due to the normal wear and tear of shipping, distribution, and handling of packages.

Process-Dependent Properties

One property controlled by the printing process is the level of adhesion of the ink to the film surface. Adhesion is enhanced by treatment of the film surface, formulation of the ink system, and drying conditions.

Equally important in preserving the integrity of the printed image is the cohesive strength of the ink system. Consequently, some ink systems undergo "cure" during drying via a chemical change such as cross-linking.

A second property controlled by the printing process is precision of registry for repeated impressions in multi-colored images. Beyond the mechanical precision maintained in the press, a significant contributor to good registration is the level and consistency of the resistance to deformation by the film when it is heated and forces are applied to pull the film through the machine. There is always some stretching of the film and the printing press is set up to compensate for this from printing station to printing station. Film tensile properties must be consistent from roll to roll to avoid constant resetting of the press controls.

Another key property of printed film is the level of ink solvents retained in the film. Incomplete removal of solvents during drying leaves a contaminant for products that are later packaged in this film. For food products, FDA regulations prescribe the amount of residual solvents that are permitted. These maximum limits are typically in the parts per million range and depend on the toxicity. When the ink solvent is compatible with the substrate film, solvent penetrates the substrate and is more difficult to remove.

Range of Application

The range of applications of printing is broad. Conditions can be found for printing almost any plastic film structure. However, the level of productivity (film speed and yields) and quality achieved varies greatly. When stiffness and strength are lower, printing becomes increasingly difficult. Shrink films, for example, being very stretchable, require special techniques and care to minimize tensions. For high productivity and high quality imaging, stiff, strong, oriented films such as OPP and PET are preferred.

FILM COATING

In earlier sections, processes were described for putting very thin coatings on films to achieve improved adhesion and for printing images. In this section the application of thicker coatings (0.1 to 0.5 mil) on films to improve properties such as heat sealing and gas barrier will be described. While similar steps are involved, the details are different since much larger quantities of material are involved: coatings to be applied and solvents or water to be removed.

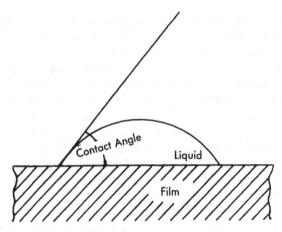

Figure 3.7 Diagram of the Contact Angle.

Film Wetting

For materials to successfully coat a plastic film, they must be in liquid form and the liquid must wet the film (that is the liquid must flow out on a film surface, such as polyethylene, like an oil rather than forming separate droplets like water). The degree of wetting is measured by the contact angle. This is the angle between the surface and a line drawn tangentially to the curvature of the droplet as shown in Figure 3.7. The smaller the angle the better the wetting. Some typical values of the contact angles for various liquids on polyethylene film are shown in Table 3.4. For any given coating system, the contact angle can be decreased by treating the plastic film or by priming as described above. In addition, the coating system can be reformulated.

Assuming good wetting is achieved, the next requirement for the coating system is to form a tough, flexible film.

Table 3.4. Contact angles for various liquids on polyethylene film.[6]

Liquid	Contact Angle (in degrees)
Water	94–102
Methylene Iodide	52
Tricresyl Phosphate	34
Hexane	0

[6]Wu, p. 138.

Coating Film Formation

Coating systems are of two kinds: dispersion, where the coating polymer is dispersed in water often using a surfactant, and solvent, where the coating polymer is dissolved in a solvent. In both systems, the liquid medium is evaporated once the coating is applied leaving behind a coating film.

Dispersion Coating

If a puddle of an emulsion of a coating polymer is allowed to stand in room air, the water slowly evaporates and the dried polymer usually forms a fragile, uncoalesced coating. However, if the melting point of the coating system is low enough or if the temperature of the puddle is raised high enough, a film will be formed. This film will be brittle or flexible depending on the degree of coalescence and the molecular weight of the polymer.

The first requirement for coalescence is that the particles must make intimate contact with each other. This will naturally occur during evaporation as the concentration of the particles increases. Diffusion and interpenetration of the polymer molecules must then occur readily across the particle interfaces. The effectiveness of this diffusion depends on the mobility of the polymer molecules which in turn depends on the temperature and the viscosity of the liquefying particle, which is a function of molecular weight. Therefore, a balance must be sought between a molecular weight low enough for coalescence but high enough for adequate toughness and flexibility of the coating film. In addition, the surfactant used to create a stable dispersion can be a barrier to interpenetration at the surface of the particles. Therefore an optimum must be found in the concentration of surfactant. While control of the factors described above will lead to a tough, well coalesced film, two more variables are involved in achieving a smooth coating: particle size and particle size distribution. Large particles tend not to coalesce completely and so produce a rough or bumpy surface. The particle size distribution should be narrow with no particles larger than the limit for a smooth surface.

Solvent Coating

Film formation from a solvent is simpler since the polymer molecules are already dissolved and naturally form a tough, flexible film provided the molecular weight is high enough.

Coating Formulation

Coating polymers are usually prepared by polymerization in water containing a surfactant so that a stable emulsion ready for dispersion coating is produced. A coating polymer to be applied in a solvent must be recovered from the dispersion by drying, washing, and drying again. The polymerization must produce the required particle size distribution. Finally, the polymer is usually made from a mixture of monomers to achieve the best balance of properties. To develop the principles here, two types of coating systems will be considered: coatings for barrier and coatings for heat-sealing.

Barrier Coatings: Measurement of Permeability

Usually, the permeation of gases through films is determined in an apparatus that maintains an atmosphere of gas pressure on one side of a film and establishes initially a vacuum on the other. The rate of increase of pressure on the vacuum side is used to calculate the rate of gas permeation or permeability. The units for permeability are typically cm^3 of gas diffused times mils of film thickness per day per 100 in^2 per atmosphere of pressure. Sophisticated instruments allow the measurement of the diffusion of specific gas molecules that is necessary where gas mixtures are of interest. An example of this need to measure mixtures is for a film whose gas permeability is sensitive to the presence of moisture so that relative humidity must be varied to quantify the effect. These instruments also detect gas concentrations at very low pressures, which is essential for measuring permeabilities for high barrier films in practical time periods.

Barrier Coatings: Permeation of Molecules through Polymers

For a small molecule such as O$_2$ or H$_2$O to permeate a plastic film, the molecule must be soluble in the film and must be able to diffuse through the film. In an amorphous film, it is easy to visualize the small molecules finding sites to occupy in the spaces between the molecules or bundles of molecules that are only loosely entangled. For diffusion to occur, the diffusing molecules must be able to move from site to site. If a polymer chain or bundle of chains is in the way, then the molecule must wait until natural vibration of the chain opens up a hole. For a polymer molecule that is relatively flexible, the movement of one part of the chain is readily accommodated by the rest of the chain and holes open rather easily. For relatively stiff and bulky polymer molecules, movement is slow and infrequent. Summarizing all this, the diffusion rate for small molecules through amorphous polymers is generally high and such polymers are poor barri-

Table 3.5. Oxygen permeability of polyethylenes.[7]

Polymer	% Crystallinity	Permeability to Oxygen [cm³-mil/(day-100 in²-atm)]
LDPE	50	480
HDPE	80	110

ers. When these polymers are made stiff and bulky (for example, poly-arylates containing large aromatic units) the barrier increases substantially.

Since any mechanism that reduces the mobility of the polymer system reduces the diffusion rate of small molecules, it follows that crystallinity is an effective way to produce a high barrier to small molecules. Polymer crystallites are so dense and ordered that small molecules can neither dissolve in nor diffuse through them. Thus, a polymer that had a completely crystalline structure would be a perfect barrier. However, such structures are not practically achievable and, for reasons to be discussed later, would not yield commercially useful coatings. Thus, all real crystalline polymers contain amorphous or disordered regions that confer permeability on the structure. In general, then, for these polymers the level of barrier is dependent on the level of crystallinity. An example of this is shown in Table 3.5 for polyethylene.

Barrier Coatings: Crystalline Polymers

In addition to crystallinity, another major factor that influences gas barrier is the specific chemical structure of the polymer molecule that affects intramolecular bonding, efficiency of packing, and rigidity. These factors not only affect the degree of crystallinity but also control the rate of permeability through the amorphous or more disordered regions. The comparison in Table 3.6 shows an example of the effect of chemical struc-

Table 3.6. Oxygen permeability of HDPE and 6,6 nylon.[7]

Polymer	% Crystallinity	Permeability to Oxygen [cm³-mil/(day-100 in²-atm)]
HDPE	80	110
6,6 Nylon	20	8

[7]Comyn, J., ed. *Polymer Permeability*. London, England: Elsevier Applied Science Publishing Co., Inc., p. 286.

Table 3.7. Oxygen permeability of selected polymers.[8,9]

Polymer	Permeability to Oxygen [cm³-mil/(day-100 in²-atm)]
Polyvinyl Alcohol (PVA)	0.01
Polyacrylonitrile (PAN)	0.04
Polyvinylidene Chloride (PVDC)	0.1
6, 6,6 Nylon	1.5–2.5
Polyethylene Terephthalate (PET)	3.0
Polyvinylidene Fluoride	4.5
Polyvinyl Chloride (PVC)	8.0
Polyvinyl Fluoride	15.0

ture. The key force at play in this comparison is hydrogen bonding, which greatly increases intermolecular attraction in nylon.

A favorable combination of crystallinity and chemical structure occurs in only a few crystalline polymers to produce truly high barriers, as shown in Table 3.7.

In the case of the first three barrier polymers, the same factors that cause high barrier also cause such high melting points that thermal degradation occurs well before melting. Therefore, the crystallinity must be lowered by introducing other monomers into these polymer chains during polymerization. The question is, can the melting point be lowered to a range that allows coatings to be formed before too much loss in barrier? The effect of comonomers on melting point and barrier is different for different crystalline polymers. Data for some typical systems is shown in Table 3.8. These are the minimum levels of comonomer content to achieve an acceptable level of processibility. PAN copolymers lose considerable barrier due to

Table 3.8. Oxygen permeabilities for various copolymers.[10]

Base Polymer	Comonomer	Permeability to Oxygen [cm³-mil/(day-100 in²-atm)]
PVA	30%	0.017
PVA	40%	0.17
PVDC	10%	0.25
PVDC	20%	0.50
PAN	30%	1.00

[8]Bakker, M., ed. 1986. *The Wiley Encyclopedia of Packaging Technology.* New York, NY: John Wiley and Sons, Inc., p. 51.

[9]Comyn, p. 285.

[10]Bakker, p. 51.

the rapid loss of crystallinity. PVA copolymers, on the other hand, become melt processible while retaining their high barrier. However, these copolymers are not generally useful for dispersion coatings and are not soluble in practical coating solvent systems. Thus, the only high barrier coating copolymer left is PVDC, which at the comonomer levels shown above yields tough films from either dispersions or solutions. At the higher comonomer concentration shown in Table 3.8, they are also melt processible. Commercially available copolymers of PVDC contain one or more of a variety of comonomers including acrylonitrile, vinyl chloride, ethyl acrylate, and butyl acrylate. Maleic or itaconic acid may also be used as comonomers to provide better adhesion.

Heat Seal Coatings

To form a heat seal on a pouch, two layers of coated film are placed between open heated jaws which are then closed and held under pressure. After a second or less time the jaws are opened and the newly formed seal cools and solidifies. The heat from the jaws must pass through the film structure to reach the coating layers which are in the center where the two films come together. Since a significant temperature differential must exist to drive enough heat for melting into the center in a second or less, the coating must melt at a temperature well below that at which the substrate film will also melt or distort. Further, upon melting, the coating layers must have a low enough viscosity to flow readily under the low pressure applied by the jaws, usually 1–2 psi. When the jaws are opened, the newly formed seal is immediately subjected to peeling forces. These forces may be slight (resistance to bending of a stiff film) to strong (sticking of the jaws to the outside surface of film or the force of a product being dumped immediately into the package). The resistance of a hot, newly formed seal to peeling is called hot tack. Even when the heat seal is allowed to completely solidify before use, there is substantial abuse during the subsequent filling, shipping, and handling.

Polymers that perform well as heat seal coatings are generally soft, low-melting, and have a molecular weight that balances reasonable melt viscosity with high solid cohesive strength and flexibility. Generally, melting should begin to occur at 100°C. The peel strength of the seal should be at least 100 g/inch and preferably 500 g/inch or more. These are typical properties of relatively noncrystalline polymers. Several of these are available as dispersions such as acrylics and polyvinyl acetate. For solvent coatings, copolyesters are used that dissolve readily in the same solvent systems used for barrier coatings.

However, it is more efficient to attain both heat sealing and barrier properties in one material. Only vinylidene chloride copolymers provide this combination in coatings, though even in this case at some compromise in barrier level. Additives such as plasticizers are also effective in obtaining the best balance of properties in these systems. For both dispersion and solvent coating, similar vinylidene chloride copolymers are used.

The Dispersion Coating Process

Dispersion coating of plastic films is similar to the printing process described earlier. The film must be unwound, the coating applied uniformly at the desired thickness, the coating dried, and the film wound up again into a uniform roll. Coating processes generally fall into two categories:

(1) *Excess application* where an excess of coating is applied to the film and the surplus removed subsequently.

(2) *Precise application* where the required amount of coating is pre-applied to an applicator that transfers it to the film.

Excess Application

Early dispersion coating processes were based on the cellophane coating operation that is shown in Figure 3.8. This coating process is typical of excess application systems. Here, coating is picked up by the film in the dip tank and the surplus is removed by the doctor rolls, which are carefully machined to yield a uniform coating thickness. Streaks or slight ridges that persist beyond these rolls are removed by the smoothing rolls. The film is dried using hot air. The rehumidification chamber is not required for plastic films. In the setup shown, film is simultaneously coated on both sides but arrangements can be made to coat only one side. Other means have been developed for removing excess coating including air knives, scraper blades, etc. All of these methods fill in the contours of the film surface so that higher areas tend to have a thinner coating. To insure a uniform barrier level throughout the sheet, excellent thickness uniformity of the base film is required.

Precise Application

By contrast, precise application systems such as the gravure coating process coat all the contours at an equal thickness and therefore are more tolerant of variations in base sheet thickness. A typical system is shown in Figure 3.9. Here surplus coating material is removed from the gravure roll before it contacts the film. Gravure systems are especially good for thinner coatings. Many modifications of this coating application process are found

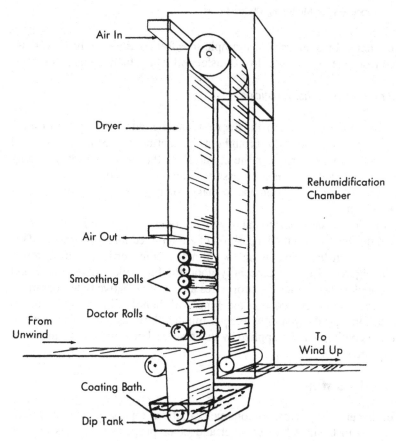

Figure 3.8 Coating Process for Cellophane.

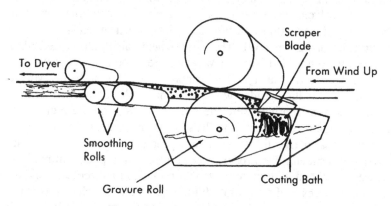

Figure 3.9 Gravure Coating Process.[11]

[11]Park, W. R. R. 1969. *Plastics Film Technology*. New York, NY: Van Nostrand Reinhold Company, p. 72.

in commercial machines. For example in offset coating, the bath is transferred from the gravure roll to a transfer roll which then contacts the film.

Other Process Considerations

In both of these processes the dispersion being fed to the applicator must be uniform in composition throughout the coating run. Storage tanks and dip tanks must be stirred or the agitation of the dispersion by pumping must be adequate. It is also necessary to analyze the coating bath during the course of the application and to make corrections to maintain constant composition.

In the drying step, horizontal dryers are often used in which the film is floated through the oven on a cushion of air supplied through nozzles. This leads to much less tension on the film than the vertical coating tower shown Figure 3.8 and is especially good for films like polyethylene and polypropylene whose tensile properties are more sensitive to temperature.

Typical commercial coating machines will handle film up to 72 inches wide at film speeds up to 500 feet per minute. Also, as in printing, several different coatings can be applied by using several coating and drying steps in sequence.

Solvent Coating

For dispersion systems, comonomer variations represent the only practical means to balance barrier level and other properties such as coating toughness and adhesion. For solvent systems, however, additional degrees of freedom are available. The solvent can assist in improving adhesion to the substrate film by solvation of the surface. Other ingredients can be dissolved in the solvent system: e.g., plasticizers to lower the melting point of the coating formulation and waxes, to increase heat sealability with minimum loss of barrier level. These ingredients also help coated films release easily from heated sealing jaws and slide smoothly over metal surfaces in packaging machinery. The formulation of coating baths evolved into a fine art over the decades when cellophane film dominated flexible packaging. Solvent coatings were the most effective, and the recipes for specific films often contained up to 20 separate ingredients.

To solvent coat plastic films, two major problems had to be overcome. The first was achieving the proper level of adhesion to plastic films. Formulas were modified with ingredients that provided the necessary chemical attractive forces to the substrate film. The second problem was static electricity generated when plastic films are unwound, especially at high speeds. This required a more complex solution.

Static Electricity in Plastic Films

The generation of static electricity is unique to nonconductors of electricity and is caused in plastic films by the separation of electrical charges on the surfaces of two films when they are pulled apart. Since there is no flow of electric current, either across the air gap or through the plastic film, the charges will not neutralize each other and very large voltages can be created. In fact, a voltage high enough to ionize air is readily reached as shown by the sparking that occurs when a charged film is passed near a conductor. This sparking must be eliminated since it can ignite the explosive mixture of solvent and air that is inevitable in any solvent coating process. While this hazard is the most serious consequence of static electricity in plastic films, there are other important problems. For example, packages charged with high static voltages tend to attract dust particles and very quickly attain a shelf-worn look. Static voltages can cause two sheets of plastic film to attract each other so strongly that it is very nearly impossible to pull them apart, as in opening a pouch in a packaging operation. When plastic is used to package electronic components, discharge of stored static electricity will damage them.

Since plastic films are repeatedly wound and unwound from rolls and passed over nonconducting surfaces at high speed, it is impossible to avoid the build up of static charges on the surface. Hence the only solution is to remove the charges just prior to solvent coating. This can be done with a variety of devices. The simplest way is to contact the film surface with a tinsel rope connected to an electrical ground. This usually provides a sufficiently intimate contact to bleed off the charges even when the film is moving at high speeds. More sophisticated devices utilize a radioactive source to ionize the air around the moving film surface to create a conductive environment that removes the charges. Even when using such devices, the level of static voltage must be monitored at various critical points in the coating sequence.

A better solution to the problem of static electricity would be to make the plastic film permanently conductive. Uncoated cellophane, with its high water content and high polarity, is sufficiently conductive to be static free. Some plastics are also conductive by virtue of fairly exotic chemical structures, but these are prohibitively expensive for packaging applications. Rendering plastic films conductive by some additive such as metallic particles requires an impractically high concentration of additive. A more practical approach is to make just the surface conductive. If film transparency is not required, this can be done by a thin metal layer such as aluminum created by vapor deposition. Where transparency is important, coatings are used that attract a conductive moisture layer. These coatings are generally used only where the final package must be antistatic, and they are not generally used as manufacturing aids for solvent coating.

The Solvent Coating Process

Except for the static elimination steps, solvent coating is very similar to dispersion coating. However, for solvent coating, the solvent must be recovered and reused since coating baths contain only about 20% solids.

The first step in reusing the solvent is to recover it from the air stream used to dry the coating. This is not easy since the solvent is at very low concentration. Usually very large carbon beds are used to absorb the solvent vapors. With heat or steam, the adsorbed vapor is later freed and then purified by distillation. When mixed solvents are used, the composition of the purified solvents must be analyzed so that the feed to the coating formulation step can be adjusted with virgin solvent to achieve the desired mixture.

As indicated above, a coating bath formula contains a variety of ingredients to achieve all the desired coated film properties. A typical formula is given below in Table 3.9.

While PVDC copolymers are dominant in solvent coating, in theory, any polymer that is soluble in a solvent could be coated onto a plastic film. In practice, given the complexity of the solvent recovery process, it is not practical to vary the solvent system and therefore only those polymers soluble in the same solvent used for PVDC coatings are used as alternatives. For copolymers of 80% vinylidene chloride or less, acetone will often

Table 3.9. Typical solvent coating formulation.[12]

Component	% by Weight	Comments
PVDC	85	80–95% vinylidene chloride plus other monomers
Plasticizer	10	Increases flexibility, lowers heat seal temperature. Phthalate, adipate and citrate are typical.
Wax	3	Improves block and slip. Carnauba, paraffin, and microcrystalline types used.
Anti-static Agents	2	Usually wetting agents
Slip Agents	<0.1	Talc or inorganic pigments
UV Absorbers	<0.1	

[12]Park, p. 66.

serve. For more crystalline polymers, tetrahydrofuran or methyl ethyl ketone (if the solution can be heated) often mixed with toluene is used.

Measurement of Coating Thickness

Given the dynamic nature of a typical coating process and the importance of coating thickness and uniformity, the coating operator cannot depend on laboratory measurements of coating thickness made on finished rolls. Coating thickness should be measured across the film as part of the coating sequence. This is done using radiation absorption as described for the cast film process. The complication here is the need to differentiate the coating from the film.

One approach is to use a form of radiation that is absorbed only by the coating. For example, PVDC coatings strongly absorb x-rays because of the high concentration of chlorine atoms. In other cases, infrared radiation at wavelengths absorbed more strongly by the coating than by the substrate film are used. As the chemical nature of the coating more nearly approaches that of the substrate, differentiation becomes increasingly difficult. One solution in such cases is to measure the substrate film alone and then the coated film and use the difference to determine the coating thickness.

Comparison of Dispersion and Solvent Coating

Process Dependent Properties

For PVDC coatings, the barrier properties of solvent and dispersion coated films are very similar and are controlled by the selection of the specific copolymer, the formulation ingredients and the coating thickness. A disadvantage for solvent coatings is that the coating is also a solvent barrier, making it very difficult to drive off the last traces of residual solvent. Since the residual solvent acts as a plasticizer, it opens up the crystalline structure enough to significantly reduce its barrier to small molecules. Additionally, these solvent residues can add unwanted taste and odor to packaged foods. All coating solvents are regulated by FDA and the process must be controlled to keep them below the allowable limits.

For PVDC coatings with heat sealing capability, solvent coating with its broader flexibility in formulation has a decided advantage. Generally, dispersion coatings are at best only marginally heat sealable. However, even with the best heat seal coatings, it is difficult to attain a hermetic seal comparable, for example, to polyethylene sealed to itself. This is partly because of an upper limit on coating thickness of about 0.2 mil (versus 1–2 mils for an extrusion-coated layer of polyethylene) and partly because of the sealing characteristics of vinylidene chloride copolymers. For similar reasons, PVDC systems do not yield heat seals with good hot tack.

Coatings used for heat seal only can be satisfactorily applied by both processes.

Surface properties of coatings applied by either process are similar and controlled largely by additives to the coating bath.

Range of Application

Dispersions can be applied to a wide range of substrate films. For films with non-polar surfaces, such as polyethylene and polypropylene, surface treatments must be employed to achieve wetting of the dispersion on the film and adequate adhesion of the dried coating. Stiffer, oriented films such as OPP and PET are ideal substrates because they remain stiff at the drying temperature for dispersion coatings. On the other hand, even the most temperature sensitive films (LDPE or PVC) can be dispersion coated commercially with very careful control of drying conditions and film tension.

Solvent coatings are used on a much narrower range of plastic films since many are sensitive to the solvents used. In fact, PET is the only plastic film that is solvent coated on a large scale.

Typical Systems

The properties of dispersion coated OPP films are summarized in Table 3.10.

The acrylic coated film with its broad heat seal range is used primarily for high-speed, overwrap applications. The PVDC coated barrier film is used primarily as a component in multilayer structures.

Table 3.10. Properties of coated OPP films.[13]

Property	Uncoated	Acrylic Coated	PVDC Coated
Haze (%)	3	3	3
Heat Seal Range (°C)	0	110–150	120–150
Oxygen Permeability [cm³-mil/(day-100 in²-atm)]	160	150	1–3

[13]Bakker, p. 323.

Table 3.11. Properties of coated PET films.[14]

Property	Dispersion Coated	Solvent Coated	
	PVDC	PVDC	Copolyester
Haze (%)	7	7	7
Oxygen Permeability [cm³-mil/(day-100 in²-atm)]	0.5	0.4	9
Heat Seal Strength (g/inch)	0	200*	250**

*Measured at 140°C.
**Measured at 120°C.

The properties of typical coated PET films are summarized in Table 3.11.

EXTRUSION COATING

In the description of the blown film process, melt drawing was discussed at some length. Melt drawing means the thinning of a film in the molten state by stretching it. For the maximum extent of drawing, the melt must, on the one hand, flow readily; but on the other hand, it must have enough strength to withstand the stretching force. In the blown film process, the demands of the bubble process and the need for final film toughness put limitations on the extent of melt drawing. However, when the polymer melt is supported as a coating on a substrate film the coating polymer can be formulated to maximize the degree of melt drawing and produce much thinner films as coatings. This approach led to the extrusion coating process which is now widely used for laying down coatings of essentially all extrudable polymers.

Polymers are tailored for extrusion coating primarily by reducing polymer molecular weight for increased melt flow. This not only increases extrusion rates but also increases the extent of melt draw. Additives, such as low molecular weight forms of the polymer or compatible waxes, are also used to increase melt flow. Increases in melt strength are attained through other structural modifications of the polymer such as, for polyethylene, the presence of a small number of very long branches.

As for any coating process, the polymer melt must wet the substrate for uniform coverage and must adhere to the substrate film. For this purpose, treatments of the substrate film surface are often required.

[14]DuPont Mylar®. *Polyester Film for Packaging—Technical Data.* 1990. Wilmington, DE: E. I. du Pont de Nemours & Co., Inc.

Extrusion Coating Process

The extrusion coating process resembles the cast film process, which uses a roll to quench the film. In the case of extrusion coating, the substrate film is passed over the quench roll or between a quench roll and a nip roll to receive the falling polymer melt. A diagram of a typical process is shown in Figure 3.10. Typical film widths for this process are 48 to 72 inches. Steps of extrusion and film formation are similar to those in the cast film process except for the effects due to the much higher speeds and extent of melt drawing for extrusion coating. Extrusion coating speeds are normally above 2000 feet per minute with some designed to operate at 3000 fpm. The highest speeds are generally achieved when coating paper. Both the speed and the extent of drawing are ultimately limited by the ability to control the edges of the melt film as well as by the strength of the melt. As draw (controlled by the differential between the speed of the substrate film and the rate of melt extrusion) is increased, the amount of neck-in and thickening at the edges increases, requiring a film die that is wider than the substrate film and the slitting off of the heavy edge of the coated film. Also, as the degree and rate of draw are increased, the edges of the melt film become increasingly unstable, leading to a scalloped pattern. Finally, as the strength of the melt is exceeded, the melt will tear

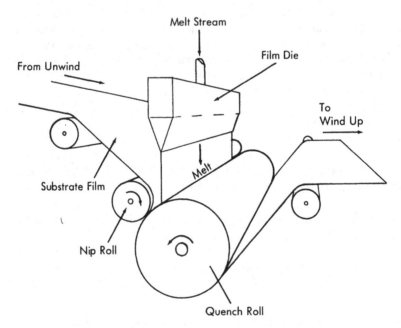

Figure 3.10 Extrusion Coating Process.

away from the die lips. Even before these problems are encountered, subtler effects such as reduced adhesion are seen due to decreased levels of oxidation.

Another difference between the extrusion coating and cast film processes is that the extrusion step tends to be at much higher temperatures for the former. For example, for polyethylene, the melt temperature for extrusion coating is 220 to 320°C, as compared to polyethylene film normally extruded at less than 200°C. For this reason, polymer formulations often include increased amounts of additives that improve resistance to thermal degradation. Even so, fumes from polymer melt degradation are a hazard and adequate ventilation must be provided around the extrusion die.

The steps of unwinding the substrate film, feeding it through the process steps, and winding it again onto a roll are essentially identical with those for printing. The principles described earlier for measuring coating thickness of coated films also apply to extrusion coating.

Process Dependent Properties

The primary variables that affect properties are the choice of coating polymer and thickness. Process variables have a secondary effect. The high extrusion temperatures not only facilitate higher extrusion rates and draw ratios but also improve adhesion because of increased oxidation of the melt in the air gap between the die and the substrate. The lower melt viscosity also aids wetting. The drawbacks of higher temperature extrusion are higher odor and lower heat seal strength as oxidation is increased.

Range of Application

The extrusion coating process was developed to coat paper or paperboard and aluminum foil. It has not been widely used for plastic films. Partly this is because the addition of layers of extrudable polymers is more economically accomplished by coextrusion, as described in a later section. Also, given the high temperature of the polymer melt, the substrate film must have high thermal stability. Typically, extrusion coating is used where it is difficult to add the layer as part of the film formation sequence as, for example, for oriented films. It is also favored where thicker coatings (about 1 mil) are desired.

Theoretically, any extrudable polymer can be melt coated onto a plastic film. However, practical considerations narrow down the field. Very high melting polymers, polymers with low melt strength, and heat sensitive polymers are ruled out. Melt coating polymers of commercial interest include the polyethylenes, ethylene copolymers, acid copolymers, ionomers, nylons, copolyesters, and recently introduced modified polyesters. Coating thickness is limited at the low end to about 0.2 mil under the best conditions.

Typical Systems

Outstanding heat sealing characteristics are the primary advantage of the thicker coatings attainable by extrusion coating. The specific characteristics are determined by the choice of coating polymer and the thickness of the coating. LDPE is the most common and lowest cost choice. Typical conditions of extrusion are a melt temperature of 220° to 320°C and a die opening of at least 20 mils. Coating thickness ranges from 0.2 to 3 mils. Polymers that are higher cost and often more difficult to process are used where enhanced functionality is desired such as a wider heat sealing range, improved sealing through contamination, increased hot tack, peel seal, and gas barrier.

The minimum sealing temperature for LDPE is about 120°C, while HDPE is even higher at about 140°C. Where lower temperature sealing is required for higher packaging machine speeds, ionomers with a minimum seal temperature of 104°C or EVA coatings that seal at 65°C are used.

When it comes to sealing through contamination such as fatty food particles or liquids, most coating materials are poor. However, acid copolymers are better and ionomers are outstanding.

LDPE has poor hot tack. Improvements are seen with LLDPE, VLDP, and acid copolymers, but ionomers are outstanding with hot tack strength at least double that of LDPE.

Packages that open by peeling rather than by tearing are desired for many applications such as pouches of snacks, bags of cereal and lids on thermoformed trays. Peel seals are achieved by weakening the normal strength of a heat seal by reducing the cohesive strength of the seal layer, the adhesion of the coating layers to the substrate or the degree of fusion during heat sealing. Generally, this weakening is accomplished using blends of the typical extrusion coating resins.

For improved barrier to moisture, HDPE with three times the barrier of LDPE is the usual choice. The choices for extrusion coatings with oxygen barrier better than LDPE are limited. Nylons and the new, modified polyester resins with about a 15-fold improvement over LDPE are most often used. Coextrusion coatings using EVOH are also being introduced.

FILM LAMINATION

Given the availability of a wide range of uncoated and coated films, their properties can be broadened considerably by combining two or more of these films with themselves or with nonplastic substrates such as paper, aluminum foil and cellophane. Thus, a manufacturer, by the installation of one piece of equipment for accomplishing such combinations, is immediately able to make a very broad range of possible packaging structures.

The Lamination Process

The key step in laminating is the creation of strong adhesive bonds between the films. All of the technologies previously described for coatings are variously utilized to lay down these adhesives which are applied from emulsions or solution. Adhesives can be either thermoplastic or thermoset. Thermoplastic adhesives, such as plasticized vinyl acetate/vinyl chloride copolymers, lack heat resistance so the lamination would be restricted in its heat sealing range. Thermoset adhesives undergo crosslinking after the lamination has been made leading to greater heat resistance. Polyesters and polyurethanes are thermosetting adhesives that contain an initiator for the curing reaction. Emulsions of acrylics are also used as is PVDC when gas barrier is needed, but they give only moderate levels of adhesion. Shown in Figure 3.11 is a typical lamination process where emulsion or solvent adhesives are used. Both wet bonding, as shown below, and dry bonding are used. In the latter case the adhesive is dried

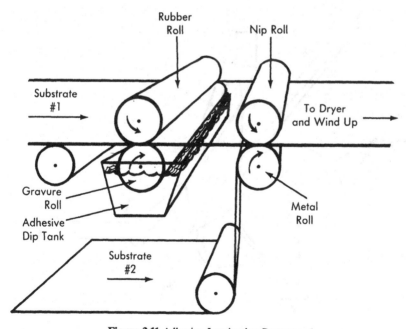

Figure 3.11 Adhesive Lamination Process.

prior to lamination. The adhesive can be applied by any of the coating processes described earlier.

Extruded adhesives are also used such as polyethylene and ethylene copolymers. High adhesive application temperatures are often employed to oxidize the melt and improve adhesion. Waxes are used to modify melt properties. A lamination process using extruded adhesives is shown in Figure 3.12.

Commercial laminating machines can handle film widths up to 80 inches and film rolls up to 50 inches in diameter. Film speeds are usually about 1000 feet per minute. Like other processes for modifying films, film must be unwound, fed through the process and wound again onto a uniform roll. Unwind and wind-up stations are often the turret type that allow ready splicing of new rolls onto the moving film, so the process runs without interruption. A complication in laminating is that two different films or a plastic film and a nonplastic substrate must be simultaneously brought together to be combined. Good mechanical and electrical precision are necessary to keep film speeds coordinated. Control of tension is critical to avoiding wrinkles.

A problem in the lamination process is the tendency for air bubbles to

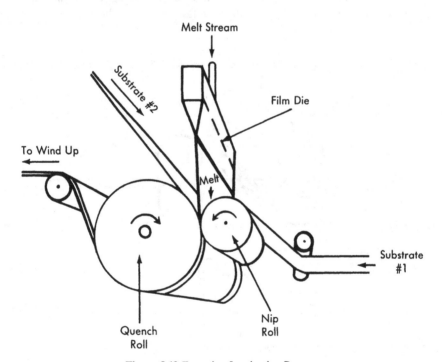

Figure 3.12 Extrusion Lamination Process.

be formed in the adhesive layer since the air layer adhering to the films is carried into the nip and trapped in the solidifying adhesive layer. The problem gets worse with increasing film speed. Increasing the temperature of the combining rolls helps because it reduces the viscosity of the adhesive layer. Another common practice is to maintain a shallow layer of liquid in the trough formed by the combining films as they feed into the nip.

When plastic films are combined with nonplastic substrates, techniques appropriate for handling these substrates must be employed. For example, thin aluminum foil, given its lack of ductility and low tear strength, requires especially careful tension control and avoidance of nicked foil edges.

Process Dependent Properties

The primary property controlled in the lamination process is the adhesion between the materials being combined. This is measured using the peeling technique described earlier in this chapter. All other properties are established by the selection of the materials to be combined. In addition to these properties, stiffness or bulk in the final lamination is usually desirable especially in applications where a very stiff package helps to protect fragile contents or resist puncture, as in industrial bags.

Another advantageous consequence of lamination is that printing on one of the plastic films can be buried within the structure. To achieve this burying the film must be reverse printed using a mirror image of the impression and in the lamination the unprinted side faces the outside, providing a glossy exterior for maximum visual appeal while protecting the printing from scuffing during shipping and handling.

Range of Application

There is virtually no limit to the combinations of materials that can be laminated together. Thermal lamination is the least versatile technique since the substrates being combined must heat seal to each other. Extrusion coating to provide an adhesive layer increases the range of substrates but the adhesive polymers that are suitable for extrusion coating are limited. The most versatile laminating sequence includes steps for treating (for example, corona discharge), steps for coating (including dispersion, solvent, and extrusion coating), and material handling equipment for plastic films, paper, and aluminum foil. Usually, however, large commercial laminating machines tend to be dedicated to a narrow range of laminations, and lack broad versatility.

Table 3.12.

Properties of a Lamination	PVDC Coated OPP/Adhesive/LDPE	
Thickness	2.9	mils
Seal Strength	4500	g/inch
Oxygen Permeability	0.9	[cm³-mil/(day-100 in²-atm)]
Tear Strength	66–230	g/inch (MD-TD)

Typical Systems

The examples[15] in this section illustrate the variety of structures that can be produced by lamination.

PVDC Coated OPP Laminated to LDPE

This structure combines bulk or stiffness with the high barrier of PVDC and the strong heat seals of LDPE. It consists of 1 mil OPP film coated with PVDC which is adhesive laminated to a 2 mil film of LDPE. Properties are summarized in Table 3.12. A similar structure could have been made by extrusion coating LDPE onto the coated OPP, although the thermally sensitive substrate would make this a more difficult operation.

PVDC Coated OPP Laminated to OPP

This example illustrates the laminating of two oriented films to gain thickness. The PVDC coating provides both barrier and heat seal. The lamination is: 1 mil OPP/adhesive/0.7 mil OPP coated with PVDC. The properties of this structure are shown in Table 3.13. Note the lower level of barrier compared to the first example above, where the coating is more crystalline since heat sealing is provided by the LDPE. The lower heat seal strength for thin PVDC coatings versus thick LDPE is apparent. Finally,

Table 3.13.

Properties of a Lamination	OPP/Adhesive/OPP Coated with PVDC	
Thickness	1.8	mils
Seal Strength	150	g/inch
Oxygen Permeability	4.0	[cm³-mil/(day-100 in²-atm)]
Tear Strength	20–11	g/inch (MD-TD)

[15]Ibid. p. 462.

Table 3.14.

Properties of a Lamination	Oriented Nylon/LDPE-Ionomer	
Thickness	2.6	mils
Seal Strength	4000	g/inch
Oxygen Permeability	9	[cm³-mil/(day-100 in²-atm)]

the low tear strength is typical of an oriented structure, whereas in the first example, the tear strength is characteristic of LDPE.

Oriented Nylon Laminated to an LDPE/Ionomer Coextruded Film

In this example 0.45 mil uniaxially oriented nylon is thermally laminated to a 2 mil coextruded film of LDPE and ionomer. Properties are shown in Table 3.14. This thermal lamination offers the modest barrier of nylon in a relatively simple structure with LDPE providing moisture barrier and the added layer of ionomer providing outstanding heat seal performance. Again, this structure could have been made by extrusion coating using the coextrusion technology to be described in a later section to provide the LDPE and ionomer layers. The choice here, as in many other cases, may be dictated by the equipment and expertise that the manufacturer happens to have rather than by any superiority of one approach over the other.

PET Film Laminated to LDPE Coated Aluminum Foil

In this last example, 0.48 mil PET film is laminated, using a 0.7 mil layer of extruded LDPE as the adhesive, to aluminum foil that had been previously extrusion coated with 2.2 mils of LDPE. The lamination is made so that the thick LDPE coating on the foil is on the outside to provide the heat seal layer. Properties of the structure are summarized in Table 3.15. Here the PET film can be printed. Note the excellent barrier provided by the foil.

Table 3.15.

Properties of a Lamination	PET Film/LDPE/Aluminum Foil-LDPE	
Thickness	2.8	mils
Seal Strength	2800	g/inch
Oxygen Permeability	<0.1	[cm³-mil/(day-100 in²-atm)]

METALLIZATION

Aluminum foil is often included in laminated structures. It adds high barrier to gases, a metallic luster for decorative purposes, and the ability to sustain a crease or fold referred to as deadfold. The first two properties can also be achieved by coating a thin layer of aluminum onto a plastic film. This is done by depositing aluminum vapor onto the film in a vacuum chamber.

Vapor Deposition of Aluminum

When aluminum is melted within a vacuum of about 10^{-2} atmospheres or lower, aluminum vapor above the liquid diffuses throughout the chamber and deposits on any surface. If a plastic film is held above the liquid, the aluminum atoms that strike the film will stick to it under the right conditions. Cooling the film with a chill roll greatly increases the efficiency of condensation. Effective cooling and short exposure times are also essential to avoid film shrinkage and distortion caused by the absorption of the kinetic energy of the depositing atoms and radiant energy from the molten pool of aluminum.

The Metallization Process

A machine for the vacuum deposition of metals onto plastic films is shown in Figure 3.13. Film is unwound from a roll, passed through the metallizing station, and wound again on a roll. While typical commercial machines can handle film a wide as 72 inches and roll diameters as large as 60 inches, newer machines will handle 96-inch wide film and operate at speeds of 2000 to 3000 feet per minute. Not only are vacuum chambers built large enough to contain such large rolls, but the vacuum pumps have the capacity to achieve a high vacuum in about 5 minutes. These modern machines are often compartmentalized so that the high vacuum of the metallizing station is isolated from sources of gas such as the air emerging from the unwinding roll. In addition to high vacuum, the other critical requirement is the ability to produce a uniform coating of aluminum across a wide film. In a typical commercial machine, the molten pool is held in a rectangular crucible. To achieve a uniform thickness across the film, the crucible length would need to match the film width. For wide films, a series of crucibles are spaced so that the deposition patterns overlap the right amount to achieve uniformity.

Several approaches are used to create the molten pools of aluminum. In

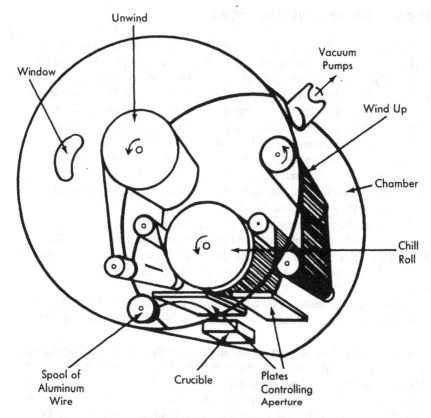

Figure 3.13 Metallization Process.

the simplest, a charge of aluminum ingot is added to the crucibles and resistance heating is used for melting. In such a system, the length of a run
is limited by the charge in the crucible. To extend run length, a wire of aluminum is fed to each crucible automatically from spools installed within
the vacuum chamber. Improved methods such as high voltage electron
beams are being developed to supply heat. These approaches increase production rate and permit the evaporation of higher melting materials.

In principle, productivity can also be increased by allowing film to enter
and exit the vacuum chamber. This permits vacuum deposition to run continuously while fresh rolls of film are spliced on a depleting roll just as in
other film processing sequences. This approach requires effective air-to-
vacuum seals that can accommodate a plastic film moving at high speed
and dragging a layer of air along with it. While successful air-to-vacuum
systems have been reported over the years, all commercial machines still
employ a totally enclosed machine.

Process Dependent Properties

The thickness of the metallized coating is controlled during the vacuum deposition process by the temperature of the aluminum pool and by film speed. Thickness is measured either by optical density (amount of light transmitted) or electrical resistance. Often these are monitored within the chamber. Both are well-defined functions of metal thickness and conversion tables have been established using standard samples of known thickness. Normally, coating thickness is in the range of 0.3 to 0.6 × 10^{-3} mil (1.3 to 2.3 optical density). For decorative purposes, a uniform metallic luster is obtained with even thinner coatings, and for high barrier, coatings thicker than 1×10^{-3} mils are needed. The relationship of metal thickness to barrier is shown in Table 3.16 for PET films.

Though coating thickness is the primary determinant of the level of barrier, deposition conditions and the nature of the film surface are claimed to be important secondary factors. Smooth film surfaces are desirable. This is not an area that is well understood and operating know-how is generally kept secret.

An important property of metallized films is that they retain their barrier properties after abuse better than aluminum foil. This is shown in Table 3.17.

Another important property is the adhesion of the metallized layer to the film. This is measured using pressure sensitive tape and the peel test described earlier. Good adhesion depends on the absence of contamination on the film surface. Common problems in this regard are oil deposited during the film formation sequence (a particular problem with films oriented in a tenter frame) and low molecular weight components exuding to the surface from within the film. Adhesion and appearance of the coating are also affected by deposition conditions. The higher the vacuum

Table 3.16. The relationship of barrier to metal thickness for metallized PET film.[16]

Metal Thickness (mils × 10^{-3})	Permeability	
	To Oxygen [cm³-mil/(day-100 in²-atm)]	To Water Vapor [g-mil/(day-100 in²)]
0.48	0.1	0.07
1.15	0.05	0.017
1.43	0.04	0.013
1.55	0.017	0.01

[16]Ibid. p. 444.

Table 3.17. The effect of abuse on the barrier properties of
metallized PET and aluminum foil.[17]

Number of Flexes*	Permeability to Oxygen [cm³-mil/(day-100 in²-atm)]	
	Foil	Metallized PET
0	0	0.06–0.09
10	0.21	0.15
100	Failed	0.27

*Gelbo flex test.

the better. Good adhesion for thicker coatings deposited at high rates is difficult to attain.

While the metal layer imparts the necessary gas barrier, other properties must be added to achieve a useful packaging structure. The ability to heat seal is imparted either by coating or by lamination to another film such as polyethylene. In these subsequent film processing steps, great care must be taken to avoid damaging the fragile aluminum layer.

Range of Application

While just about any plastic film can be metallized using vapor deposition, the demands of the process when running at high rates limit commercial operations to only a few films. Specific requirements of the film are high strength and thermal resistance. Also, films are preferred that do not have volatile additives that either migrate to the surface and/or interfere with the achievement of a high vacuum. As a result, most metallized films are based on PET, OPP, and nylon.

Given their outstanding gas barrier, metallized films would seem to be a better choice than polymer coatings such as PVDC. However, aluminum coatings have their limitations. They are opaque whereas many applications require transparency. Aluminum metal is a barrier to microwave energy and is unsuitable for microwaveable packages except as susceptors (see Chapter Six). Further, the metal is fragile and non-ductile, disqualifying it from the many packages that are made by thermoforming the film into a shape.

Typical Systems

The level of barrier achieved by metallization is a function of the substrate barrier as well as the thickness of the metal layer. Shown in Table

[17]Skodis, L. C. 1989. "Films, Metallized" *Packaging Encyclopedia*. p. 21.

. *Table 3.18. The effect of the substrate on the barrier of metallized films.*[18]

Film	Permeability to Oxygen [cm³-mil/(day-100 in²-atm)]	
	Unmetallized	Metallized (2.3 OD)*
LDPE	420	10
OPP	80	4
PET	5	0.08
Nylon	2.7	0.07

*Optical density of 2.3 = 0.6 × 10⁻³ mil thickness.

3.18 are data for several films with the same thickness of metal. Metallization can be thought of as an enhancement of the natural barrier properties of the substrate. Note in Table 3.18 that even the poor gas barrier of polyethylene film can be improved about 40-fold to approach the moderate barrier of films like PET. OPP, already with an excellent moisture barrier, can also gain an oxygen barrier like that of PET by metallization. In fact, at thicknesses of metal greater than 1×10^{-3} mil, metallized OPP has a barrier approaching uncreased aluminum foil. The same is true of PET. Oriented nylon, on the other hand, starts with a good gas barrier but poor moisture barrier. Metallization raises its moisture barrier to that of OPP, making it suitable for packaging moist products.

FILM SLITTING

For high volume commercial film operations, the route to increased productivity comes via wider and wider machines. Widths of 100 inches are common for cast and blown film lines whereas 200 inches is characteristic of the newest oriented film lines. On the other hand, widths of machines for modifying films described in this chapter are often 72 inches or less. Film widths for packaging machines are typically in the range of 12 to 36 inches. Therefore, film manufacturers are required to offer film of almost any width, often in increments of 1/4 inch, so that film slitting is an important processing step.

A typical film slitter is shown in Figure 3.14. The hardware for unwinding from a roll, feeding film through the slitting step and winding again onto a roll is similar to that described earlier for printing and other film processing sequences. The principles for tension control and for achieving uniform rolls are also common to other processes.

In Figure 3.14, film is slit using razor blades where the film passes over

[18]Ibid. p. 444.

a roll with grooves and the razor blades are positioned part way into the grooves for precise control of the cut. For many films, the edge of the razor blades are quickly dulled at the point of contact. The razor blade is therefore oscillated back and forth so that the full length of the edge is used. For films with low tear strength like oriented films, maintenance of a sharp blade or knife edge is critical in avoiding a nicked or torn edge from which a tear will propagate across the sheet. Another common method is shear slitting whereby two circular sharp edges in contact provide a cutting action like a pair of scissors. Sets of these knives are positioned across the sheet using holders that can be precisely set and easily repositioned. Shear slitting is especially recommended for films that are tough or thick. Recently, slitting with high-speed water jets or laser beams has been introduced. Both approaches claim the advantages of higher speed slitting and the generation of tear-resistant slit edges. For laser beam slitting there is the additional advantage that the slit position is easily changed by shifting the position of the mirrors that control the path of the beam. However, the added complexities are not usually justified in the routine slitting of packaging films.

Once the film has been slit, each new segment must be separately wound onto its own roll. Thus the windup consists of a series of rolls containing

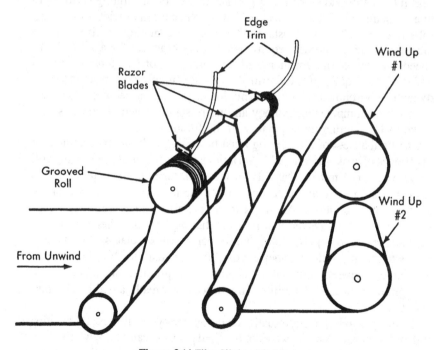

Figure 3.14 Film Slitting Machine.

cores cut to the width of each sheet. In Figure 3.14, two center winding rolls are shown. The winding cores can also all be placed onto a single windup roll using discs for separation and control of each winding roll. For surface-wound rolls each winding roll is often supported by a set of arms that cause the rolls to press down onto a common drive roll.

Machines are available for slitting films as wide as 80 inches at speeds up to 1500 feet per minute, producing final slit rolls with complete flexibility in widths and maximum diameters of 24 inches. Special machines have been designed for even wider films, but given the massiveness of these machines and the resulting increase in the time required to change rolls and slitting widths, manufacturers often prefer to use these large machines to prepare rolls in widths convenient for transfer to smaller machines which produce the final slit rolls.

The addition of process computers and more sophisticated electronics to the most recent machines has improved productivity by allowing quick changing of the widths of the slit films and of the winding conditions. The slitting program can now be changed without shutting down the operation.

Like any other film processing step, slitting can be added to the film formation or film processing sequences where it is done just prior to winding the film onto a roll. Good isolation of the tensions from the rest of the process must be provided to insure precise control of the slitting. Slitting in line is usually more economical only when the pattern of widths of the slit rolls remains relatively constant. The frequent changing of slit widths to meet the varied customer requirements is not practical for a wide, high-speed film operation. Furthermore, the precision of film tension control and tracking required for uniform slit rolls is difficult to achieve under the dynamic environment of a film formation or processing sequence. The application of computer technology and more sophisticated electronics may make in-line slitting more popular in the future.

One final aspect of film slitting must be considered: the management of the process for maximum yield. Variations in the width of the starting roll are at best very limited. Thus, the task is to fit together the many different widths of slit rolls that are needed so that the maximum utilization of the starting roll is achieved. At one extreme, the array of slit widths can consist of the orders received that hour or that day. At the other extreme, if delivery lead times permit, the orders can be accumulated over days or weeks to maximize the number of possible combinations. If order patterns are stable, frequently ordered slit widths can be produced ahead for inventory giving increased flexibility. This process is greatly aided by computerization.

To understand the continually changing array of film widths being ordered, consider that the width of the final package can be anywhere from a few inches to 36 inches but is typically between 6 and 20 inches. Thus, the printer will order film widths depending on his orders in hand and the number of package widths he can fit into his press width. The printer then orders the optimum width for his schedule of jobs from, for example, a film laminator who goes through the same process of width optimization and so on down the line to the orders placed on the primary film producer. In all of this sequence, coordination between the various producers of machine widths for maximum width utilization has not been practical.

ECONOMICS

The approach used here will be the same as in the discussion of the economics for the various film formation processes. However, here it is not practical to define unit investment since the processes are being carried out on substrate films that can vary greatly in thickness. Here the manufacturer is more interested in the speed at which films can be run through the operation. Therefore, investment quoted will be for hardware to process a standard film width of 72 inches at speeds achievable with standard equipment. The manufacturing cost indicated will be the cost added by the processing step to the cost of a standard 0.5 mil oriented film.

Investment

The investment for the pieces of equipment needed to carry out the various processing steps discussed in this chapter is difficult to determine, since much of the hardware is custom built and since several steps are often combined. However, as a starting point, slitters represent the simplest hardware for processing films and are readily available in standard designs. Their cost ranges from $150,000 to $250,000. This investment can be used for the unwinding, film handling, tension controlling, and winding functions that are common to all film processing hardware. Adding to these common functions a step like treating or more complex steps such as priming and printing, where drying is also required, increases the investment to the range of 1.5 to 2 times the slitter investment. For example, a printing press with three-color capability costs from $400,000 to $600,000. For coaters, much more extensive drying and air handling are required and, when solvents are used, recovery and purification steps must be added. These systems are largely custom designed, leading to a wide range of investment, but the lower end of the range would be 2 to 3 times that of a printing press. Extrusion coaters are comparable in investment to cast film lines (about $800,000 to $1,200,000) and laminators cost more because of the added cost of drying and solvent recovery, depending on the adhesives used. Finally, a metallizer, which adds the complication of a high vacuum chamber, costs about $1 million but can range up to double this with the latest technology for highest film speeds and minimum pump-down times.

Manufacturing Cost

Material Cost

Any processing step causes some wastage of the substrate film. Even though film yields are generally quite high (95% or better) in film processing, material costs can be significantly increased by this wastage especially where high-cost oriented films are involved. For example, for oriented films that cost from $1.00 to $2.00 per pound, each 1% loss adds $0.01 to $0.02 per pound to the finished product. Coating materials can add from a few cents/lb for thin coatings, as in priming, printing, and metallizing, to as much as $0.20 per pound of film for thicker coatings of higher-cost materials such as PVDC. In solvent coating, solvent losses are negligible in the recovery and purification systems so that the added cost is insignificant. However, where solvents are incinerated, the added cost for solvent can be as much as that of the coating material. In extrusion coatings, where much thicker coatings are used, the weight of added material is often several times that of the substrate film. Thus, the added cost can be as much as $0.90 per pound of processed film for a 1 mil coating of polyethylene on 0.5 mil PET or OPP. Laminations are still more costly since they are 3 to 6 times thicker than monofilms and they are built up by combining purchased films.

Operating Cost

The operating cost for a simple process for modifying a film is similar to that for a cast film or blown film line. Therefore, for operations that run continuously with minimum down time for changeovers, and at maximum film widths, the operating cost is in the range of $0.05 to $0.10 per pound of 1 mil product. However, most of these processes are not run at these optimum conditions and the operating cost varies widely—up to 2 to 3 times this minimum. Labor efficiencies are usually improved by moving crews among several operations as orders demand and by having machines of different width capabilities.

COEXTRUSION

We have indicated from time to time that it is often more economical to add layers during the original film formation process. Thus, dispersion, solvent and extrusion coatings have been added to commercial film lines.

A major advance was made when the process of coextrusion was developed. Here two or more streams of different polymer melts are joined in the melt film formation step in such a way that the resulting layers maintain their separate identities in the final film. With the advances made in recent years in extrusion technology and in polymers designed specifically for coextrusion, this technique is becoming widely used to create multilayered films, as part of the film formation sequence, that match the combination of properties achieved heretofore only through a sequence of separate processing steps.

There are two different approaches to producing a coextruded melt film: (1) combining melt streams in the die body and (2) combining them outside the film die using a feed block.

The Design of Coextrusion Film Dies

In a coextrusion die, separate channels for flow are machined into the die body for two or more streams and the streams join just prior to emergence from die lips. A typical flat film die for two layers is shown in Figure 3.15. The merging point for the two separate streams is chosen based on several considerations. In general, the longer the two streams are

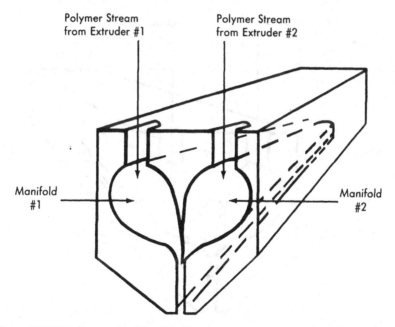

Figure 3.15 Flat Coextrusion Film Die (Center Fed) Cross-sectioned through the Feed Ports.

in contact in the molten stage, the better the adhesion between layers in the final film. On the other hand, if the streams differ widely in viscosity, merging close to the die lips is preferred. The latter approach is also essential where one of the streams is much less resistant to thermal degradation than the other. In some cases, elaborate die designs have been created to allow separate tailoring of the flow path and conditions for each stream. For example, the flow channel for the less stable stream can be kept much shorter and thermally insulated from the other. In some cases the streams are merged just as they both emerge from the die lips.

These same principles have been applied to circular film dies as well. These pose additional complexities since in circular dies, good mixing in the flow channels is required to eliminate the weld lines formed by the merging of the polymer streams from the different feed ports in the die body. If two different polymers are used, each must pass through its own spiral section and the final merging of these different polymer melts must occur very near the die lips. If the circular die is rotated to achieve randomization of thickness variations, the complexity is further increased.

The coextrusion dies described so far are relatively inflexible since flow channels must be machined into the die body. A significant increase in flexibility is gained with choke bars or vanes which adjust the relative flow

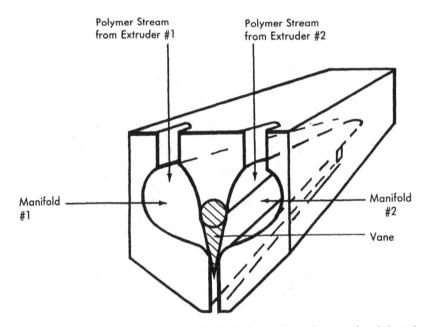

Figure 3.16 Flat Coextrusion Film Die (Center Fed) with Vane. Cross-sectioned through Feed Ports.

rates of the streams just prior to their merging. These are controlled from outside the die body allowing adjustments on the fly. An example of a vane system is shown in Figure 3.16. Such a die is much more complex and costly to machine especially as the number of polymer streams is increased above two.

Merging the polymer streams exterior to the die body is a much lower cost and even more flexible approach; however, for this to succeed, stable laminar flow must be maintained within the die so that the various streams maintain their separate identifies and relative thicknesses. This stable flow is achieved by streamlining the die cavities and is shown for a center-fed, flat die in Figure 3.17.

The Design of Coextrusion Feed Blocks

Given the capability for stable laminar flow in the die, a feed block can be inserted just upstream of the die to accomplish the merging of different polymer streams. An example of a simple feed block is shown in Figure 3.18. This system is essentially a miniature multilayer sheet die. The number of layers is predetermined by the number of flow channels. In more sophisticated feed blocks, vanes are added which allow adjustment from outside the block of the relative flow volumes of each stream (Figure 3.18b). Also distribution pins, adjusted as well from the outside of the block, are placed at the merging point. These permit adjustment to compensate for the inherent effect of the die when streams differ greatly in thickness or viscosity. With these added features, layer thicknesses are more readily changed and differences in viscosity as high as 40:1 can be handled with distribution differences across the sheet of $\pm 3\%$. In practise, most flat coextrusion film processes use the coextrusion feed block and a single manifold die. For circular die systems, both external and internal combining of melt streams are utilized.

With these advances, the burden of development has been shifted from the hardware to the design of polymers to achieve the necessary compatibility in the combinations desired.

Polymer Systems for Coextrusion

While advances in equipment have greatly increased the tolerance for the coextrusion process to handle large differences in viscosity, it is still desirable to match the flow characteristics of the different polymers as closely as possible. This is hard enough at a given shear rate and temperature, but the fact that the viscosities of different polymers respond differently to shear rate and temperature makes the problem of matching even more difficult. As a result, it is necessary to think of a coextrusion as

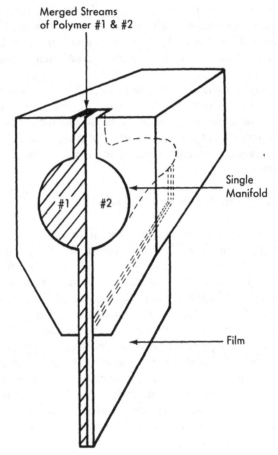

Merged Streams
of Polymer #1 & #2

#1   #2

Single
Manifold

Film

Figure 3.17 Flat Film Die with Two Stable Polymer Streams Cross-sectioned through Center Feed Port.

a polymer system with each component specifically designed to be part of the system. Typical coextrusions offer combinations of benefits, analogous to coated or laminated structures, that are achieved by selecting layers for stiffness and strength, for heat sealing, for barrier, and for achieving adhesion between chemically unlike layers. In addition, a layer may be included for recycling film waste generated through edge trimming and losses during start up and transitions.

Coextrusions for Heat Seal

For a relatively simple coextruded structure, such as a film with a surface layer for heat sealing, little modification of the polymers is required

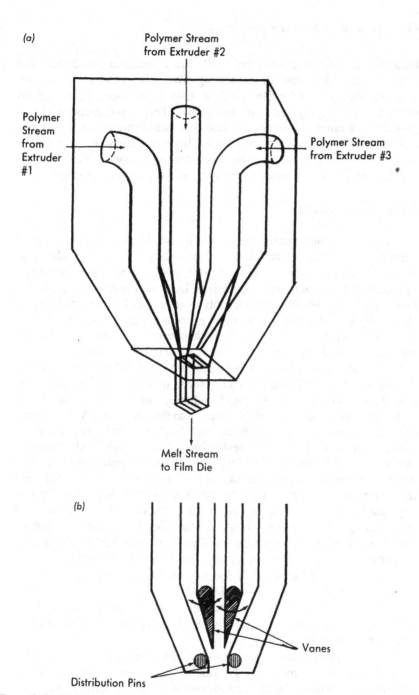

(a)

Polymer Stream
from Extruder #2

Polymer
Stream
from
Extruder
#1

Polymer Stream
from Extruder #3

Melt Stream
to Film Die

(b)

Vanes

Distribution Pins

Figure 3.18[19] (a) Coextrusion Feed Block. (b) Diagram of Added Vanes and Distribution Pins.

[19]Rauwendaal, C. 1986. *Polymer Extrusion*. Munich, Germany: Hanser Publishers, p. 454.

for a process to be successful. For OPP films, copolymers of ethylene and propylene are often used for the sealing layer. Alternative seal layers for this and other polyolefin films include EVA copolymers, EAA copolymers, EMMA copolymers, and ionomers. These copolymers provide a broader seal range, an increasing ability to seal through contamination, and, especially for ionomers, outstanding hot tack. Adhesion of the sealing layer to the substrate is attained by adjustment of the comonomer ratio and the flow properties are matched by adjustment of molecular weight.

Barrier Coextrusions

Complexity is introduced when a gas barrier is also required. Typical barrier polymers are crystalline and more difficult to extrude without excessive thermal degradation. Furthermore, modification of flow properties by changing the comonomer composition leads very quickly to loss of the level of barrier. Finally, barrier polymers are unlike the other layers chemically so that achieving adhesion to the other layers requires the use of separate adhesive layers. Thus, both barrier and adhesive polymers must be designed specifically for use in a coextruded system.

The commonly used barrier polymers are nylon, PVDC, and EVOH. Both nylon and EVOH are relatively easy to process with the other components in a coextrusion system. In addition, their thermal stability is sufficient to allow recycling of film waste into the structure. PVDC is more difficult to process, and a relatively high level of comonomer content is needed to make it suitable for coextrusion. Thus the high levels of barrier that are typical for the more crystalline copolymers applied by solvent or dispersion coating are compromised in the extrudable versions. Even with this compromise, extrusion temperatures are still close to the thermal degradation threshold, making the control of the coextrusion process and reprocessing of waste more difficult. On the other hand, nylon and EVOH have limitations in barrier properties. Shown in Table 3.19 are the barrier properties of the polymers commonly used in coextruded structures as a function of relative humidity. Nylon has only a moderate barrier to oxygen, which is reduced at higher relative humidities, so it must be used in fairly thick layers. EVOH, while excellent at low humidity, rapidly loses barrier at relative humidities exceeding 80%. PVDC shows a constant barrier with increasing humidity.

Thus, it is apparent that an ideal extrudable barrier polymer does not yet exist. EVOH is the most widely used in coextrusions because of the relative ease of processing. The sensitivity of the EVOH barrier to humidity can be offset by design of the coextruded structure. For example, the barrier layer can be protected from the ambient humidity by using a heavy layer of a good moisture barrier such as HDPE. Where the package con-

Table 3.19. The effect of relative humidity on the barrier properties of EVOH, Nylon and PVDC.[20]

Barrier	Relative Humidity	Permeability to Oxygen [cm³-mil/(day-100 in²-atm)]
EVOH (30% Ethylene)	0%	0.017
	65%	0.02
	95%	>25
6 Nylon	0%	1.5
	95%	5.0
PVDC (80% Vinylidene Chloride)	0%	0.5
	95%	0.5

tents are the source of moisture, the structure is designed with a low moisture barrier on the outside to permit rapid passage of moisture to the air to avoid buildup in the barrier layer.

Coextrudable Adhesives

Coextrudable adhesives are now available for attaching together almost any combination of layers, as Table 3.20 shows. In Table 3.20 the X's in-

Table 3.20. The adhesion of coextrudable adhesives to various resin types.[21]

	Coextrudable Adhesive							
	Anhydride Modified				Carbonyl Modified	Acid modified		
Resin Type	PE	PP	EA	EVA	EVA	EVA	EA	EVA/EA
Polyethylenes	X		X	X		X	X	X
PP		X	X			X	X	X
PET			X					
Ionomer	X			X		X	X	X
EVOH	X	X	X	X				
Nylon	X	X	X	X			X	X
PAN					X			
Acrylics					X			
PVC					X	X		
PVDC					X	X		
Polystyrene					X	X		

[20]Ibid. p. 51.
[21]Fischer-Drowos, S. G. 1989. "DuPont Bynel® Coextrudable Adhesive: Product Line Update". Presentation (February). Private Communication.

dicate where there is adequate adhesion for most applications. Adhesives that can be used for a particular combination are found by locating the columns that have X's for each resin type in the combination. For example, good choices to adhere EVOH to a polyethylene resin are anhydride-modified PE and EVA.

The Coextrusion Process

Except for the extrusion and melt film formation steps, the basic processes for cast, blown, flat oriented, or tubular oriented films are unchanged by the addition of coextrusion. The die must be altered and/or a feed block added as described above and a separate extruder must be used for each different polymer fed to the process. The size of each extruder is generally set to provide a range of throughputs appropriate for the relative thickness of the layer it supplies. Sometimes one extruder will feed more than one layer when they contain a common polymer such as adhesive layers on either side of a barrier layer to adhere it to polypropylene on the one side and a heat seal layer on the other.

Reuse of waste film is more complex for coextrusions since the different polymers used are often too incompatible to be mixed with any of the pure layers. In many cases a separate recycle layer is added which is supplied by a separate extruder.

Most coextrusion systems operate in the range of 200 to 1,000 pph. A typical three extruder system would use 3.5- and 2.5-inch extruders although many 4.5-inch extruders are used with some lines operating as high as 2,000 pph with 6-inch extruders.

Process Dependent Properties

The properties of a coextruded film are a straightforward combination of those contributed by each layer. Thus, selection of the polymers for each layer establishes the properties of the final film. Coextrusion leads to no unique synergism except occasionally interlayer adhesion is enhanced because the long times during which the melt streams are in contact. Given the selection of polymers for each layer, further control of properties is attained by their relative thickness.

Under the best conditions, layer thickness uniformity in the 1% range is attainable across the web. As might be expected, the measurement of the thickness of each layer is difficult. Typically, however, each layer has a sufficiently different index of refraction that it can be readily distinguished when looking at a cross section of the structure in a microscope. A scale can be superimposed onto this cross section and the thicknesses determined. This is a tedious process, however, so it is usually not practical to make the many measurements that would be required to assure uniformity of the layer thicknesses in the MD and TD. The development of devices for on-line measurement of layer thicknesses is receiving much attention in the industry. Already available are infrared systems using several wavelengths, each strongly absorbed only by a different layer. These have capability for two to three layers. Newer technology will provide additional wavelengths. PVDC layers are a problem since they exhibit no strong infrared bands and separate gamma ray systems must be employed for them.

The precise control of layer thickness is most critical for barrier layers since the relatively high cost of these polymers dictates using the minimum essential thickness. This makes uniformity essential if each package made from a roll of film is to provide the required shelf life for its contents. For lower-cost heat sealing polymers, extra thickness to assure performance is not a serious cost penalty. For adhesive layers, the thinnest possible coverage is usually adequate. To achieve a high precision in the control of layer thickness, all the principles outlined for the cast film process must be applied for constancy of flow from the extruder, for achieving uniformity across the film as it is formed in the die, and for the control of film tensions that affect the degree of melt draw. This again emphasizes the importance of designing each component polymer of the system so that, as much as possible, flow and melt draw characteristics are similar, including the response of these characteristics to temperature and shear rate.

Range of Application

Coextrusion can be added to any film formation sequence or film processing sequence where extrusion is employed. However, a recent survey showed[22] that many still consider coextrusion an art rather than a science, with many improvements in hardware and materials still needed. For barrier structures, this survey found that laminations are still slightly preferred over coextrusions with metallized structures close behind the two. However, coextrusion is more and more commonly used in the cast film and tubular film processes. For oriented films, given the complexities of the orientation process, it is less common. In fact, for anything but the simplest oriented systems, it is generally found to be more economical to use separate film processing steps such as solvent coating or extrusion coating. This is because the yield for orientation sequences is only in the 55 to 75% range, so that a large amount of trim and waste film must be recycled. The amount of waste from coextruded films that can be recycled is generally limited due the incompatibility of the different components. This problem can be reduced by designing the coextrusion die so that the

[22]Ericksen, G. 1989. *Packaging*. 34, p. 47.

multilayers are not extended into the beaded edge which is later trimmed off.

Coextrusion is also being used more and more in extrusion coating. This permits the addition of an adhesive layer in addition to a seal layer and increases the range of possible heat seal coatings. The addition of an adhesive layer also has opened up the ability to add barrier coatings.

Typical Systems

OPP/Ethylene-Propylene Copolymer

This is one of the few examples of an oriented coextrusion produced in large volume. It is typically about 1 mil in total thickness, including a layer of copolymer on each side of the OPP layer. Heat seal characteristics for the structure are

Heat seal range	88–149°C
Heat seal strength	700 g/inch

This is a much better heat seal than the 200 g/inch level that is available from dispersion coatings of PVDC. While it lacks the gas barrier of a PVDC coating, the natural moisture barrier of polypropylene makes the coextrusion well suited for many packages. The heat seal range can be further broadened and the strength improved somewhat by moving to more highly modified copolymers and to ionomers, but none of these achieve the seal strength (about 4500 g/inch) of thick layers of polyethylene incorporated by lamination or extrusion coating. On the other hand, the advantage of the ethylene-propylene copolymer as the seal layer is its greater compatibility for blending with the substrate, which is critical because of the high level of recycle required for an oriented film process.

PP/Adh/EVOH/Adh/LLDPE

This is a good example of a barrier coextrusion that is made by the cast or blown film sequences. The adhesive layers are usually anhydride-grafted polypropylene. Typical layer thicknesses are: 0.8 mil PP/0.6 mil EVOH/2 mils LLDPE. This structure provides an excellent combination of low oxygen permeability [0.03 cm³-mil/(day-100 m²-atm), high heat seal strength (4500 g/inch) and the stiffness and good moisture barrier of polypropylene. For applications where the demands on the heat seal are lower, the LLDPE layer thickness can be considerably reduced. It is noteworthy that while EVOH draws adequately for the blown process, versions that orient well are not yet available.

The principal advantage of a coextrusion such as this over a lamination with similar properties is cost. The main disadvantage is that printing cannot be buried in a coextrusion. While considerable progress has been made in improving the abrasion resistance of surface printed images, most packagers who need a high quality print job choose laminations.

Economics

Investment

In general, the investment for extruders is about 50% higher for coextrusion than it is for a single layer system with the same output. For example, a 6-inch extruder costs about $500,000 compared to a four extruder (with diameters of 3-1/2, 3-1/2, 2-1/2 and 2-1/2 inches) coextrusion system at $770,000. The output of each system is 2,000 pph. Looking at the investment for a complete film line the increase is about 20% for coextrusion. For example, a coextrusion, blown film line producing 300 pph of film costs about $400,000. This means the unit investment is $1,333 per pound as compared to about $1,100 per pound on average for a monofilm line.

Manufacturing Cost

Operating costs are very nearly the same for monolayer and coextrusion lines. In one example, the cost was $0.064 per pound for a 6-inch monolayer extrusion line and $0.068 for the coextrusion line.[23] Thus, in general, the cost of separate lamination or coating steps is saved when a coextrusion can be substituted, and this saving is about $0.05–0.10 per pound for each step eliminated.

Summary

The advantage of coextrusion resides in process economics: for a modest investment over and above that for a film formation process many different structures can be created for essentially no increase in operating costs. Coextrusion does not lead to films with any unique properties compared to the same structure produced by other processes such as lamination or coating.

BIBLIOGRAPHY

Bakker, M., ed. 1986. *The Wiley Encyclopedia of Packaging Technology.* New York, NY: John Wiley and Sons, Inc.

Briston, J. 1983. *Plastics Films. Second Edition.* Essex, England: Longman Group Limited.

[23]Benning, C. J. 1983. *Plastics Films for Packaging.* Lancaster, PA: Technomic Publishing Co., Inc., p. 147.

Cloeren, P. J. "Coextrusion Dies to Process 9-Layers at Widely Different Temperatures", *Proceedings of the Fifth Annual International Coextrusion Conference and Exhibition 1985*; and "Performance and Economics of Barrier Coextrusion Coating", *Proceedings of the Third Annual Conference on Coextrusion Markets and Technology, October 3–4, 1983.*

McKelvey, J. M. 1962. *Polymer Processing.* New York, NY: John Wiley and Sons, Inc.

Park, W. R. R. 1969. *Plastics Film Technology.* New York, NY: Van Nostrand Reinhold Company.

Wu, S. 1982. *Polymer Interface and Adhesion.* New York, NY: Marcel Dekker Inc.

CHAPTER 4
Package Functions and Requirements

INTRODUCTION

Packaging began when early man needed some way of carrying items from one place to another. Large solid objects could be carried unpackaged, but liquids and powders required some form of container. Thus began the first and most fundamental function of the package: *containment* of a product. Since some of these packaged items were not consumed all at once, some way of dispensing part of the contents and preserving the rest for later use was needed, leading to the second and third basic package functions: *dispensing* and *preservation/protection*.

The other major package functions appeared when commercial transactions began. The sale of a product was facilitated by the use of a package to *standardize* the amount of product sold. If several identical items were sold in a single transaction, packaging provided a way of *assembling* these products. In modern marketing, *display* and *communication* have become highly important functions.

This chapter discusses the various functions of today's packages and the requirements these functions impose on the packaging material.

PACKAGE FUNCTIONS

Containment and Dispensing

The nature of these two primary functions of a package is so obvious that no further elaboration is required. The requirements they impose on the packaging material are covered in the next section.

Preservation/Protection

Once the functions of containment and dispensing are satisfied, preservation and protection are the most important functions performed by pack-

131

aging. The package contents determine the degree of protection required. All packaged items must be protected against the intrusion of foreign matter such as dirt, dust, and rain and against physical damage by outside forces. More delicate products such as food need additional protection against environmental factors such as gases and light. Some food products need a package that will maintain a special atmosphere. Retail products are subject to pilfering or tampering; a package can reduce this hazard.

Common packaging materials – glass, metal, paper and plastics – can all be used to perform all these functions to some degree, but plastic films perform all these functions save one as well as and frequently better than the other commonly used packaging materials. The one exception is protection against physical damage. Although special plastic films that contain air pockets or air entrapped in foam structures are often used for cushioning, physical impact protection is usually provided by paper in the form of corrugated cartons, balled-up newspaper, or by small "noodles" made from foamed plastics.

Dirt, Dust, and Rain

All plastic films are completely impervious to intrusion of these unwanted materials, providing protection against them indefinitely while still keeping the package contents completely visible. No other inexpensive flexible packaging material performs this function as well as plastics.

Atmospheric Gases

Many packaged food products must be protected against moisture gain or loss. Some must also be isolated from oxygen which causes spoilage from oxidation or bacterial action. For centuries, metal and glass have been used to perform this function, but the enormous weight savings afforded by flexible packaging have led to the use of this technique to package many sensitive food products. Without plastic films, which are the most inexpensive flexible materials that can provide satisfactory protection against oxygen and moisture, most food products would still be packaged in heavy glass or metal containers and many more would not be packaged at all.

In today's industrialized society, packaged food products must also be protected against man-made atmospheric gases such as hydrocarbons from automobile and truck exhausts. Many foods will absorb such gases to the detriment of their flavor and must be protected against this possibility.

The shelf life of some food products can be significantly extended if the package is flushed with inert gases such as nitrogen or CO_2 to remove air from the package headspace. When this technique, called "Modified Atmosphere Packaging" (MAP) is used, the packaging material must have low permeability to oxygen to prevent its re-entry into the package.

Closely related to MAP is the newer and more complex technique of Controlled Atmosphere Packaging (CAP). Fresh produce is often sold unpackaged, but the shelf life of these perishable products can be extended many fold if they are surrounded by a container that is selectively permeable to oxygen, CO_2, and ethylene. As discussed more fully in Chapter 6, if the permeability of the container to these three gases is within certain ranges (which differ for different produce items), the atmosphere within the container will slowly change and remain optimum for preservation of the food within. Use of this highly specialized technique for extending the shelf life of certain foods is still in its infancy, but it will become more common in the future. Plastics are the only flexible packaging materials that can be designed to meet these specialized gas permeability requirements.

The common but crude approach to providing permeability for produce that must respire is to perforate the film. This approach sacrifices gas barrier, so the maintenance of a controlled atmosphere is impossible with this technique.

Loss of Flavor/Odor

Many food products and other items such as perfume are intentionally flavored or possess odors that are essential to the product function. Just as the package must protect the contents against acquisition of undesirable odors or flavors, it must protect the contents against the loss of desirable flavor or odor characteristics.

Light

Protection against light is required for food and drug products that are sensitive to UV-catalyzed oxidation. Non-food products rarely need this kind of protection. Metal and paper, being opaque to light, automatically provide this function. As described below, plastics must be modified for this purpose.

Extremes of Temperature

Food products must sometimes be refrigerated to extend shelf life and most are exposed to ambient winter temperatures during distribution. Thus flexible food packages must not embrittle at low temperatures. All common plastic film materials have sufficient low temperature flexibility to meet this requirement, but cellophane does not. This drawback was an important reason for the replacement of cellophane by plastics in food packaging.

A few food products are designed to be heated or cooked in the package. In this case, the plastic film must resist softening or melting at cooking temperatures and must not release any of its chemical components to the contents during the process. This is a far more difficult requirement for most plastics to meet than the low temperature requirement noted above, but as discussed below and in Chapter 6, a few plastic films are safe at cooking temperatures.

Infestation

Food is attractive to insects, rodents, and other animals. One obvious but frequently overlooked package function is to protect the contents against infestation by these organisms and against contamination by other unwanted species such as airborne spores and bacteria. Almost all packaging materials that are themselves free of microorganisms and that can furnish a tightly closable package will exclude tiny species, but rats, mice, squirrels and other larger animals can often be thwarted only by metal or glass containers.

Tampering and Pilfering

Assembling small items in a large package makes surreptitious pilfering more difficult. This approach has been used for years to discourage shoplifting. Tampering, a more recent phenomenon, has concentrated on products that are ingested such as foods and pharmaceuticals. Occasionally tampering is evilly motivated, but often the tamperer just wants to sample the product. Regardless of motivation, most pharmaceutical manufacturers and some food processors now incorporate tamper evident features on or in their packages. When tampering with evil intent first began, some manufacturers attempted to create tamper resistant packages. They soon learned that a determined tamperer can penetrate any package that the consumer can also open, and so turned their efforts to tamper evidence, relying on purchasers to protect themselves by examining a package that is designed to make tampering apparent.

Standardization

The sale of many items requires that the purchaser know how much is being purchased, unless that is already obvious from the nature of the item, such as an automobile tire. The package serves the vital function of providing a standard amount of product, so that package filling can be divorced from the purchase of the packaged item.

Assembly

The manufacturer of a product frequently needs to assemble and bundle together several identical items or items that are different but are all needed to complete the product offering. In retail selling, the common "six-pack" may be the best known example of the former; assembly of a board game consisting of a playing board and a box of counters, dice and instructions is one example of the latter. Many products are bundled together for shipment on pallets or packed in corrugated or wood boxes or huge steel containers for deck transport on merchant vessels.

Display and Communication

These functions are quite different from the protective functions described above, but can occasionally be accomplished in conjunction with those other functions. An obvious example of this synergy is the potato chip bag which must be opacified to protect the chips against UV-catalyzed oxidation. Opacifying a plastic bag by pigmenting or metallizing also provides an excellent background for the elaborate graphics that some of these packages carry. The graphics look better when printed on an opacified bag, thus satisfying the need for an eye-catching shelf display. Normally however, the display and communication functions require independent techniques that do not contribute to product protection.

Advertising

Today, the supermarket and other consumer goods outlets that rely on self-service dominate retailing of packaged goods in the U.S. and in other advanced countries. With no salesperson available to influence the buyer, and with many essentially identical products competing for the consumer's dollar, franc, or yen, manufacturers must rely on package appearance and the visual messages it conveys to sell their products. Glossy packages have more eye appeal than dull ones. Artistic graphics rendered in several colors can suggest attractive features of the product that cannot be quickly conveyed in print.

An interesting statistic shows how important and effective the package advertising function can be: because so many people see it, the cost of advertising on a package, expressed in terms of cost per thousand impressions *delivered*, is less than 10 cents, compared to $3–4 per thousand impressions for a full page newspaper ad or about $20 per thousand impressions for 1/4 hour of TV time.[1]

---

[1]Hanlon. 1984. *Handbook of Package Engineering*, New York, NY: McGraw-Hill, p. 18.

Product Information

In addition to stating what the product is, information needed for correct use of the packaged product is usually printed on the package: directions for use, lists of ingredients, recipes that include the product, and so forth. Bar codes for automated scanning and billing at the checkout counter must be on the package. The manufacturer's name and the brand name are often crucial: Campbells Soup, Heinz pickles, Schlitz beer—in many cases the producer's reputation makes the sale.

Product Concealment

Normally the product manufacturer wants the package to reveal the product to the greatest extent possible, consistent with the overriding need to protect it. There are cases however where products are deliberately hidden by the package. Brittle salty snacks which tend to break into small pieces during shipment are frequently packaged in opaque films to disguise the degree of breakage. Many products are rather dull: flour, for example, is better packaged in opaque paper decorated with a picture of a smiling housewife or a field of golden grain. Some products are downright unattractive: moist dog food, for example, or beef stew. Products of this kind are usually found in packages with attractive images that conceal the appearance of the contents.

PACKAGE REQUIREMENTS

The introduction to this chapter listed the major package functions: containment, dispensing, preservation/protection, standardization, assembly and display. The requirements that these functions impose on plastic packaging films will be discussed in this section.

Containment

For a package to successfully contain a product, it must first meet certain mechanical requirements. Most important, the packaging material must be strong enough to hold the load. Strength is a catchall term for what are actually four separate properties: resistance to *breaking* when slowly pulled (tensile strength), resistance to *breaking under sudden impact* (impact strength), resistance to *puncturing* by sharp objects, and resistance to *propagation of a tear* that originates at a puncture. For plastic films, these properties do not necessarily go hand in hand. Polystyrene films have good tensile strength but low elongation which leads to low puncture resistance and impact strength. LLDPE film, weaker in tension than PS, has better resistance to impact and puncture due to its ability to yield and stretch before it breaks. This combination makes LLDPE a stronger film. Table A.1 in the Appendix gives quantitative values for these various strength characteristics of plastic films.

Given the proper mechanical properties, the packaging material must resist both chemical and physical attack by the contents of the package. Chemical attack could come from a variety of sources, ranging from oils, fats, and grease to acids and caustic liquids. Plastics vary in their resistance to attack by these agents, but a few plastic films are as inert as glass to all these substances. Physical attack by package contents usually occurs from sharp-edged products. Some food products and many hardware items pose this challenge. A moderately thick (1–2 mil) film will usually resist penetration by sharp-edged package contents unless the product is very heavy, but the common practice of packaging 25-pound blocks of ice in EVA films clearly shows that some inexpensive plastic films can handle heavy, sharp-edged products as well.

Dispensing

The essential function of dispensing means that the package be readily openable to give access to the contents. From one point of view, all plastic film packages easily satisfy this requirement so long as the consumer has a knife or a pair of scissors at hand. Packagers now realize that consumers want more than this: the ability to easily open the package without such tools. In addition, consumers want the opened package to be reclosable to contain and preserve the unconsumed contents.

These are difficult requirements for plastic film packages to satisfy. The seals on food packages must usually resist penetration by oxygen and water vapor. This important food protection requirement must take precedence over easy openability. Nevertheless, converters and resin manufacturers are now developing resin blends and special seal layers that will satisfy both the protection and the easy access requirements, but the generally higher cost of this feature will slow its adoption by many food packagers.

The nature of blister, skin, and shrink wrapped packages make them notoriously difficult to open without a sharp tool. Tear strips are incorporated in some shrink wrap systems, but this approach is not totally satisfactory from an esthetic point of view because the tear strip does not shrink along with the shrink film. This leads to puckering of the film near the tape. In addition, the higher cost of this approach deters packagers. Blister and skin packages often provide a perforated area on the back of the card that can be penetrated without tools. In sum, the whole issue of easy openability has not been resolved to the satisfaction of purchasers and remains their most frequent complaint about flexible plastic packages.

Reclosability is an even more difficult requirement for plastic film packagers to satisfy. The original seal, even if it were delicately opened, cannot be recreated without heat. Various types of zipper-like closure systems are found on some plastic bags, but these are not hermetic closures, so the product must either be able to tolerate this failing or the package must also incorporate an additional hermetic seal—a costly and awkward arrangement. Rigid containers fitted with screw caps or snap-on caps are clearly superior in reclosability to plastic film packages.

Preservation/Protection

Dirt and Dust

Since all plastic films are totally impervious to dirt and dust, this is an easy requirement for them to satisfy. Some films, however, are exceedingly poor conductors of electricity and tend to readily acquire small static electrical charges that attract dust. This can be ameliorated by treating the film or choosing a more conductive plastic.

Atmospheric Gases

The data in Table A.1 show that plastic films differ widely in their permeability to oxygen and water vapor. Oxygen permeability ranges from 0.01 cc-mil/100 in²-day-atm for EVOH to about 800 for LDPE. Water vapor permeability ranges from 0.05 g-mil/100 in²-day for PVDC to 2.8 for PVC. Thus packagers can choose the permeability they desire. By settling for high permeability values they sometimes gain other desirable properties, achieve lower costs or both. The usual practice, however, is to use a multilayer film, described in detail in Chapter 3. One film layer can perform the gas impermeability function while other layers provide strength, heat sealability, opacity, etc. Some materials, such as PVDC, provide excellent atmospheric protection along with good heat sealability. Metallized layers are excellent gas barriers and also impart opacity and esthetic appeal.

Design of a multilayer package is complex, since packagers have many different needs. Thus the purpose of this discussion is only to point out that the gas protection requirement can almost always be met by choosing the right plastic resin. The phrase "almost always" must be used because there are certain food products for which high barrier plastic film packages have not yet been adopted. These are the so-called "shelf stable" products such as soups and vegetables which are processed, canned, and last for years on the shelf without refrigeration. A plastic pouch that incorporates either aluminum foil or PVDC as an oxygen barrier was developed many years ago for this application but is used in the U.S. today mainly by the Army, because its light weight is an important plus for soldiers who must carry their food. The limited popularity of this package with civilian consumers does not lessen the impact of what this package proved: that it is possible to create an all-plastic package that will safely contain oxygen sensitive foods for long periods of time. Widespread penetration of this market by plastic awaits the successful development of the all-plastic rigid can rather than consumer acceptance of the flexible plastic pouch.

As noted above, MAP packaging requires a barrier film that will prevent oxygen re-entry. Nylon, PVDC coated films, or coextrusions that incorporate oxygen barrier polymers such as EVOH or PVDC are all used to meet this requirement. CAP packaging requires the package to be *selectively permeable* to gases when used to extend the shelf life of fresh produce that must respire. This more complex requirement can be achieved today with the right combination of plastic films. This makes for a relatively expensive package compared to a simple polyethylene bag which only contains the product and keeps it clean. For this reason, as well as others involving distribution networks and product seasonality, CAP produce packaging has not achieved widespread adoption. As time passes, however, produce packaging in sophisticated, multilayer engineered plastic films will become increasingly popular. The savings in product spoilage and the elimination of seasonal product availability are driving forces too powerful to resist.

Flavor/Odor Loss or Gain

Earlier in this chapter the need for protection of package contents against acquisition of unwanted flavors or odors from the environment and against loss of desirable flavors or odors was discussed. Plastic films that have good barrier to oxygen and water vapor will generally also protect products from flavor/odor loss or gain. Only rarely will the package designer need to incorporate special plastics in the package to perform this function.

Light

Unlike paper or metal, plastic films must be pigmented or metallized to develop opacity. This can be an advantage for plastics, since, as noted in Chapter 3, metallized coatings can be controlled to produce selective UV screening or any degree of opacity that the product requires, whereas naturally opaque materials cannot be adjusted in this way. In some cases, partial opacity can provide adequate light protection coupled with attractive decorative effects.

Extremes of Temperature

Brittleness at refrigerator temperatures is not a problem for the common plastic packaging films and so need not be included on the package designer's checklist. Stability at high temperatures is another matter. The upper temperature limit required in food packaging is about 400°F. Only PET, nylon and the less commonly used PC will withstand this temperature without loss of mechanical properties and without releasing undesirable chemical constituents to the package contents. The package designer who contemplates a "cook-in" package must be extremely careful that the choice of film be in accordance with extensive FDA regulations that deal with this matter. To stay within FDA guidelines, the choice for applications of this kind is usually limited to PET or nylon.

Tampering and Pilfering

Some plastic film packaging techniques offer what is arguably the most effective evidence of tampering. A plastic shrink film overwrap, unlike a non-shrink film overwrap, is extremely difficult to replace without the specialized equipment that was required to install it on the original package. Even if a determined tamperer were to acquire such equipment, the use of printed film by the original packager would make this approach unworkable. Although some manufacturers have adopted shrink film overwrap for tamper evidence as well as appearance and moisture protection, most use shrink bands to secure the package top or lid to the package body. Plastic film membranes, sometimes metallized or laminated to foil, are another commonly used but easily penetrated tamper resistant device. Since the latter two approaches use far less film, they are less expensive, which accounts for their greater popularity.

Package Inertness

One important requirement of any packaging material is that it be inert to the package contents. There are three separate requirements that must be met for a packaging material to qualify as inert:

(1) It must *resist attack* by the package contents which would destroy package integrity.

(2) It must not *contribute* any foreign material to the package contents.

(3) It must not *abstract* any component from the package contents.

A few plastics do not effectively resist grease and oil. Others are attacked by certain chemicals. Nevertheless, with the right film, a manufacturer can use plastic to package almost any substance. For non-food and non-drug products, extraction of package components is never an issue since that happens only in the case of liquid products, and there the amount of material extracted, if any, is so small as to be meaningless.

Food packaging is a different situation. Here the issue of consumer safety is a primary concern for the packager. Resins from which food packaging films are produced must be manufactured to specifications that ensure that traces of catalyst or polymerization by-products are kept at levels below which migration of these unwanted materials into the food could alter the taste or be harmful to the consumer. In the U.S., any new packaging resin or film is carefully scrutinized by the Food and Drug Administration. Only after their review of the manufacturer's test data satisfies them that the new material is safe do they permit its use. Until the FDA is satisfied on this point, the use of the material will either be prohibited or restricted to package layers that do not directly contact the food.

This careful FDA review has occasionally caused food packaging material manufacturers to alter their plans. The soft drink bottle based on polyacrylonitrile was kept off the market until the producer could prove that the traces of residual acrylonitrile monomer would not be a hazard to health. By the time the FDA was satisfied on this point, the PET bottle so dominated the market that the PAN-based bottle could not successfully compete with it. When evidence of the carcinogenicity of vinyl chloride appeared, PVC producers modified their polymerization process to sharply reduce the levels of residual vinyl chloride in their product. Even so, doubts about the safety of PVC are causing some food packagers, mainly in Europe, to adopt alternate plastic films.

The natural taste of some food products can be perceptibly altered by changes in the levels of certain trace components. For example, the taste of orange juice will change if certain so-called "fresh notes" are removed. Polyolefins will abstract these components from orange juice, requiring producers striving for a fresh taste to use more expensive polymers as the contact layer in plastic orange juice containers.

Standardization

Standardization is an easy requirement to satisfy, since any package, flexible or rigid, can be designed to hold a measured amount of product.

Assembly

Assembly of many items into a single package runs the gamut from huge container loads on ocean freighters to multi-packing of small individually packaged food items. Plastic films can be made strong enough, the main requirement, to assemble loads that weigh up to several tons.

For retail product assembly, plastics are unique in providing transparency, an important marketing advantage. Shrink wrapping offers superior eye appeal and lower cost by eliminating the folds and tucks of conventional overwrap. As described in Chapter 6, shrink and stretch films are widely used to bundle products on pallets for shipping. These films are strong enough to effectively immobilize the pallet load at low cost.

Communication and Display

To serve these essential package functions, a plastic film must meet several requirements. The importance of each will vary depending on the packager's marketing strategy.

(1) The film must readily accept printing inks. If the film surface can be treated so that the ink cannot readily be rubbed off, that is a plus, but is not an essential requirement: the widespread use of multilayer films allows the converter to print one of the inside surfaces so the ink is protected from abrasion by at least one layer of clear film.

(2) The plastic resin must accept and not be damaged by dyes and pigments that are used to impart color (dye) or opacity (pigment). However, since opacity can also be achieved by surface printing rather than pigmentation, sometimes at lower cost, this is not a requirement in many cases.

(3) The film must have the optical characteristics that packagers consider essential for the display or the concealment of their products. Table A.1 lists the optical properties that are important for optimum product display: gloss, haze, and percent light transmitted. Haze is a measure of the "milkiness" of a film and is caused by light scattering by surface imperfections or film inhomogeneities such as large crystallites, incompletely dissolved additives, voids, or cross linking. A haze-free film imparts a high degree of visibility or legibility to the printing on or the details of the packaged item.

Gloss measures the ability of a film to specularly reflect incident light as does a mirror. High gloss films impart a sparkling appearance to a package. Percent light transmitted is the ratio of light transmitted with the film interposed between light source and light receptor to that transmitted without the film present.

(4) The package must look good on the shelf. This is a catchall requirement, but nevertheless a very important one for retail packaged goods. The optical characteristics of the film are not the only factors that determine how the package will look. The modulus (stiffness) of the film determines how well it will stand up, rather than sag, on the shelf; its crease resistance enables it to appear unwrinkled after being

handled and then replaced by the shopper. Packagers frequently must make plastic packages thicker than required for simple product protection in order to keep them attractive.

This entire list of film requirements cannot be regarded as essential for all packages. Many products are not displayed standing up; many products need not be visible; some products can be successfully marketed in unprinted films. However, a careful inspection of packaged retail items will reveal that at least one of the requirements on this list, and frequently several, are as important to the packager as are the other requirements outlined in this section.

Cost

The cost of packaging materials is always a major concern for packagers, since these costs can be an important fraction of their total product costs. For example, producers of bagged salty snacks spend about 20% of their total product dollars on packaging.[2] Table 4.1 shows the importance of packaging costs for other selected consumer products.[3]

Costs have two main components: the cost of the packaging film and the cost of the process required to convert that film to a package and fill it with product. The runnability of the film on the packaging equipment will influence the latter cost, sometimes substantially, and that is why the issues involved in runnability discussed in Chapter 5 are so important.

Packagers are constantly involved in a balancing act between package requirements and package cost. They are always interested in minimizing cost, but the product requirements must come first. If the product is not adequately protected, it may deteriorate or be damaged, a risk few pack-

Table 4.1. Packaging costs as a percentage of manufacturer's total costs for various products.

Product	Percentage
Aerosols	80
Bar Soap	50
Beer	27
Potato Chips	17
Women's Hosiery	10
Cigarettes	3

[2] 1988. *Packaging* (January):58.
[3] Hanlon, p. 15.

agers are willing to take. Many scholarly articles have been written, notably by Labuza, et. al.,[4] on mathematical processes for calculating minimum thicknesses of film required to protect packaged products against oxygen and water vapor. By using techniques of this kind, packagers can minimize the cost of providing adequate atmospheric barrier. However, protection against many other intrusive possibilities such as tampering may also be needed, and no handy formulas exist to help the packager with this problem or with the other protection factors discussed above; judgement is the only technique that can be used.

Once protection has been provided at minimum cost, the more complicated and subtle question of package appearance must be confronted. Packagers differ widely in their devotion to package appearance. Producers of similar products such as potato chips will likely spend more money on this factor than producers who feel that their products are unique or at least well differentiated from competition.

The frequent use of aluminum foil in flexible packages is a good example of how the cost/appearance equation is tipped in favor of appearance. Even though the freshness of most packaged foods can be adequately maintained with an all-plastic structure, consumers associate freshness with a foil package. While foil is the best moisture and oxygen barrier for some foods, it is frequently used where its cost is not technically justifiable but where the manufacturer thinks that its freshness image will help to sell the product.

The marketing specialists in consumer goods companies have an important voice in package appearance decisions. This often frustrates the flexible packaging converter salesperson who has a less expensive but equally functional package concept to offer: after touting the many advantages of the new package, including its lower cost, the salesperson may be confronted with the negative argument that "our package works and customers recognize and like it—we're not willing to risk a change."

Conversely, the converter who has the skills to produce a much better looking package may often get the business even though the new package offers no functional advantages over the one in current use.

In some packaging situations, cost is such a dominant factor that it far outweighs appearance. Grocery sacks are a good example of this situation. When special grades of HDPE with extraordinarily high tensile and tear strength were developed, the thickness of bags made from this new polymer could be reduced by a few thousandths of an inch, yielding savings large enough to allow the new plastic to quickly gain a significant share of this very large market. Commodity food products such as rice,

---

[4]Labuza, T. P. 1981. "Prediction of Moisture Protection Requirements for Foods", *Cereal Foods World*, 26:335.

sugar, and flour are other examples of cases where cost rather than appearance dominates the packager's choice. Large flour producers, when offered a shrink film overwrap that provided a much cleaner, better looking package at a cost of a few additional cents per package declined the offer because of the slightly higher cost involved.

Thus it is difficult to draw any general conclusions about this last package requirement except the very general one that cost is *always* important.

CHAPTER 5

Packaging Machinery

INTRODUCTION

The machines that convert plastic resins into films have been described in Chapters 2 and 3. This chapter will describe the machines that are used to make plastic packages and to package products in plastic films.

This will be a very selective treatment of this subject, focused solely on machines designed to package products in flexible plastic films and more particularly on the components of those machines that handle the film itself and therefore exert some influence over the choice and design of those films. For example, this discussion will not deal with product feed and conveying devices, labeling equipment, check-weighing processes, cartoning, palletizing and the many other essential components of a complete product packaging line. This limitation allows concentration on the important *film* characteristics that are required by the machines.

Altering Packaging Machines for Plastics

Equipment for flexible packaging originated in the era when paper and paper-based products like glassine and cellophane were the only flexible packaging materials available. When plastic films appeared in the early 1950s, packagers and machinery manufacturers were forced to alter their equipment to adapt to the different characteristics of these new films. Today, some of the older flexible packaging machines in operation are hybrids retrofitted to run plastics, while newer equipment has been specifically designed to handle plastic films.

From the machinery point of view, the major differences between plastic films and paper-based films are stiffness, stretchability, and heat sealability.

Stiffness

Few plastic films can compare in stiffness with paper or cellophane. PET, the stiffest commonly used plastic film, is only 80% as stiff as cellophane, while LDPE, the limpest plastic film, has only 5% of cellophane's stiffness. Since it is easier to design and operate a machine that pushes rather than pulls the packaging film through the machine, early machines were designed that way; major alterations were required to enable them to pull the much limper plastic films.

Stretchability

With their lower elastic modulus, as compared to paper or cellophane, plastic films will stretch more readily. As a result, the film tensioning devices on packaging machines had to be altered by adding pre-feed equipment that provided a loop of film prior to the main film drive. Even with this modification, plastic films are still more difficult to run than paper. This problem is particularly acute on the wrapping machines described later in this chapter.

Heat Sealability

Early flexible packages made of paper were sealed with glue-based adhesives applied to the web on a separate machine or machine station. The later development of heat sealing adhesives based on nitrocellulose or wax was a major advance in flexible packaging, since these adhesives could be used to achieve stronger, more reliable seals with some hermetic character and enabled the packager to eliminate the annoyance of solvents essential to glue systems. The still later development of plastics led to two related improvements in heat sealing: (1) the use of plastic resins as replacements for nitrocellulose and waxes as heat seal coatings for flexible packaging substrates such as paper, foil and cellophane, and (2) the elimination in some applications of heat seal coatings entirely; some plastic films have the right melting characteristics to produce satisfactory heat seals even when uncoated.

Plastic resins provide seals with much greater strength, toughness, reliability, and hermetic character than the nitrocellulose and wax heat seal materials that they quickly replaced. Although modern versions of these adhesives are still occasionally used in flexible packaging constructions, plastic resins now dominate this packaging application.

Sealing with heat required new technology and major design changes for machines that had previously run only glue-based adhesives. Additional modifications were needed when plastic films and plastic heat sealing resins appeared. Control of the temperature of the heat sealing bars is far more critical for plastics than it is for cellophane or paper, since for most plastics a relatively narrow temperature range exists within which the seal can be satisfactorily made without badly distorting the heat sensitive plastic. Cellophane or paper coated with a plastic heat seal layer require no special surface on the heat sealing bars, but when all-plastic films appeared, they softened at heat sealing temperatures and not only stuck well to themselves but also stuck to the bars, which then had to be modified with non-stick temperature resistant materials such as Teflon® to prevent this.

CHAPTER ORGANIZATION

In this chapter, the discussion of packaging machines will be divided into six sections, each dealing with a different category of package contents. The first section covers finely divided, free-flowing dry products such as powders, small particles, or flakes. The second covers liquid products. The third section covers individually wrapped solid products such as blocks of cheese, loaves of bread, or packages of cigarettes. The fourth section covers small solid products such as bolts and nuts that are not individually wrapped but are contained in a package incorporating many identical units. The fifth section discusses various heat sealing processes and the sixth section summarizes the interactions of plastic films and packaging machinery.

DRY POWDERS AND PARTICLES

The packager of dry powders and particles can choose between two fundamentally different packaging methods. In the first method, the package is created on the same machine that fills it with product. In the second method, the package in the form of a bag or pouch is made on a separate machine, often in a separate plant by another manufacturer, and brought to the machine which fills it with product and closes the open end.

Each method has its advantages and drawbacks, but generally the first is the more efficient. It is used for packaging dry, free-flowing food products such as salty snacks, beans, rice, dry soup mixes, etc. and also for liquids and pastes. This method is widely used in the food, pharmaceutical, chemical and hardware industries.

As the quantity and weight of product in the bag increases, as is the case with industrial products such as cement or plastic resins, the second method becomes more popular because it more reliably avoids the risk of the bag breaking along the bottom seam during the filling operation.

®Teflon is a registered trademark of E. I. du Pont de Nemours & Co., Inc.

Vertical Form-Fill-Seal Packaging Machines

The machine that makes and fills packages in a single operation by the first method is called a vertical form-fill-seal (VFFS) machine (see Figure 5.1 on page 153).

These machines form pouches from a web of flat, flexible heat seal material, fill them with product, and seal them in one continuous operation. The pouches may be used as the primary package or may be inserted into cartons for additional protection during warehousing and shipping.

Two materials are simultaneously fed to this machine: the film that forms the pouch and the product the pouch will contain. The film is fed from a roll to a device that forms it into a tube around the product filling tube. The two film edges are then sealed together. As the tube moves down the machine, two horizontal sealing bars come together to form a heat seal which becomes the bottom of the pouch. At that point, a measured amount of product is allowed to flow through the product filling tube and into the just-formed pouch. By the time filling is complete, the top of the pouch has traveled down to the sealing bar location, where the bars once again meet to simultaneously create the top seal and the bottom seal of the next pouch above. The sealing bars are equipped with a knife which cuts through the seal to separate the filled pouch from the machine.

In most VFFS machines, the bars that make the cross-seals contain a Nichrome wire through which an intermittent electric current passes to heat the bar to the correct temperature. Both this temperature and the dwell time must be carefully controlled to supply just enough heat at just the right temperature long enough to melt the plastic but not overheat it. Overheating can cause unsightly puckering or eventual burn-through of the film.

The side seam seal can be made either in the fin style:

Fin/Foldover Seal

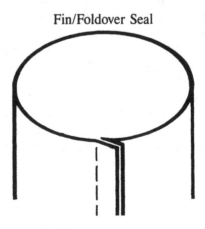

or the lap style:

Lap Seal

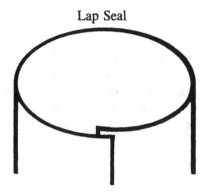

The latter consumes less film and is somewhat more attractive, but requires the seal layer to be applied to both sides of the film. The fin seal is usually used where hermetic seals are essential or in cases where there is the likelihood that the product will sift through small gaps in the seal.

The pouch made as described above is called a pillow pouch. A flat bottomed pouch can be made in a very similar way if the machine is equipped with devices that can form four creases in the infeed film prior to making the side seam and then tuck, fold, and seal the bottom flap under the bag. Flat bottom bags are used when the packager wants a package that will stand up by itself on the store shelves. Many food items such as sugar, flour, cookies, and snacks are packaged in flat bottom VFFS pouches.

Like all packaging machines, necessary adjuncts to these VFFS machines are devices that control product flow so that an accurately known quantity of product is placed in the bag. Post-packaging check-weigh stations are sometimes used as an additional quality control measure. The multiplicity of devices used for these essential purposes will not be described in this chapter since they are not relevant to the subject of plastic films.

Film Requirements for VFFS Machines

Packagers are constantly striving to run their machines faster so they can expand output without purchasing more machines. This puts constant pressure on the manufacturers of packaging films and resins to create improved products that will allow increased machine speeds. The film requirements described below for VFFS machines also apply in varying degrees to other styles of packaging machines.

The film characteristics that largely determine how well a given film will run on VFFS and other packaging machines are coefficient of friction (COF), tensile strength, flexibility and sealing properties.

Coefficient of Friction

COF can be reduced by choosing films made from resins that naturally provide good slip characteristics or by applying coatings or additives to films that do not have good slip characteristics. For example, OPP film, which combines good strength and clarity at low cost, has a rather high COF of 0.4 which can be cut in half with acrylic coatings or with slip additives such as stearates.

Strength and Flexibility

These properties are important because on a VFFS machine, the film is pulled around the forming horn at very acute angles, as shown in Figure 5.1. Low strength and/or flexibility will cause frequent film breaks at this point in the operation.

Sealing Properties

Hot tack is probably the ultimate rate-limiting film factor on VFFS machines. As noted in Chapter 2, hot tack is a term that denotes the degree of strength, or resistance to peeling apart, that a sealant layer can develop while it is still hot after just being released by the sealing bars. Widely used packaging films such as LDPE have relatively poor hot tack. Low-cost ethylene copolymer films such as EVA have better hot tack than LDPE, but ionomer films have the highest hot tack of any plastic films used in packaging. Many packagers are willing to pay more to have these higher cost resins incorporated in their film structures because the dollar savings from the higher machine speeds attainable with these structures more than offset the higher film cost. This becomes increasingly important as the weight of the package contents increases.

Since the heat seals on VFFS machines are made very rapidly, usually with less than one second dwell time, it is difficult to maintain precise temperature control in the seal area. This puts a premium on the *seal temperature range*. This range is bounded on the low side by the temperatures below which no seal is made and on the high side by temperatures that distort or destroy the film. If this range is only a few degrees wide, the film must be modified, usually with coatings, to increase it to at least 25°F. For example, OPP, which is widely used on VFFS equipment, has a very narrow heat seal temperature range and must be coated to be runnable on VFFS equipment. LDPE or EVA coatings will increase the range to

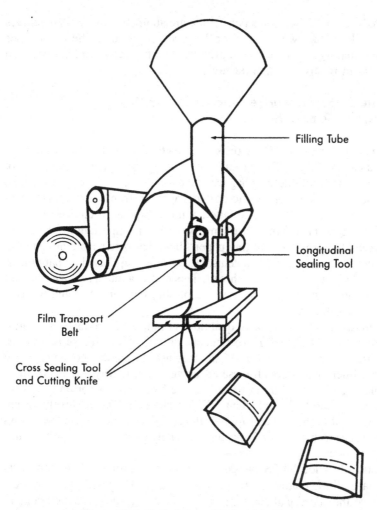

Figure 5.1 A Vertical Form-Fill-Seal Machine.

greater than 25°, but ionomers are much more effective, producing heat seal ranges up to 100°.

As the seal cools and hardens, it slowly develops its ultimate *strength*. A strong seal is vital, since pouches for dry products must withstand the frequently severe mechanical stresses of distribution and handling. Most commonly used polyolefin films will meet this requirement, but hot tack and ultimate seal strength generally go hand in hand, so EVA or ionomer films are often used in the film construction to provide this necessary characteristic as well.

Speeds up to 120 packages per minute are attainable on VFFS machines that are filling bags with up to 1 or 2 pounds of product. These machines are used mainly to package food products that typically are high cost and thus sold in relatively small quantities.

Plastic Bags for Large Quantities of Dry Industrial Products

As noted above, the VFFS packaging method is widely used for relatively small packages. The packager who wants to put (say) 50 pounds of dry free-flowing product in a bag usually chooses a bag made on a separate machine from paper or plastic-coated paper or from a combination of paper and a plastic liner. The latter, a more expensive choice, must be used if the contents are moisture sensitive or if it is important to avoid contaminating the product with tiny paper fibers that result from abrasion of the paper by the particulate contents. Heavy duty bags based on paper usually are closed by sewing with thread. This makes a much stronger closure than can be achieved with heat sealed plastic, but cannot match plastics' gas impermeability.

All-plastic bags to hold large quantities of industrial products are beginning to appear. They offer excellent moisture barrier, grease resistance, transparency, chemical resistance and inertness to product contents. Since a plastic liner also offers all these advantages save clarity, the major driving force for the adoption of the all-plastic bag is the opportunity to take advantage of the highly efficient VFFS technique. The relatively recent availability of high hot tack, high seal strength sealant resins now allows the packager to use a VFFS machine to safely package 50 to 100 pounds of dry product.

Plastic films for VFFS packaging of large quantities of dry products must have all the characteristics described above for small bags but must also have the additional strength and toughness needed to safely contain the much greater product weight. LLDPE, with tensile strength at least twice that of LDPE, provides these necessary features along with high seal strength, good hot tack, and the ability to be sealed well even when the seal area is contaminated with the powdered contents of the bag. At about the same cost, HDPE films possess the necessary strength and toughness, but are inferior to LLDPE in clarity and seal characteristics. Some dry products such as certain plastic resins require an even greater degree of moisture protection. For these, higher cost laminations or coextruded structures incorporating barrier layers such as 6 nylon are now available, but expensive structures of this kind are likely to be used only when the high cost of the bagged product can justify their adoption.

Pouch or Bag Machines

Plastic bags come in a variety of styles and sizes, but can most easily be classified by the way the film is sealed to produce the bag.

Side Weld Bags

This is the simplest film bag. It is formed by folding the web of film and making two seals transverse to the film direction. The fold creates a bottom and the two side welds create the rest of the bag. The top is left open for later filling. The bags are cut apart by a knife that is an integral part of the downstream sealing head. Small side weld bags are used for small hardware items, larger ones for frozen foods, and even larger ones for garbage.

For larger or thicker products such as loaves of bread, a gusset is created in the bottom by a gusseting attachment on the bag machine. This allows the bag to be opened and stood on its bottom to receive the contents. The sandwich bag is another special type of side weld bag that contains a flap that is folded down and sealed into the side welds, leaving an opening through which the sandwich can be inserted, after which the flap is folded over to make the closure. The part of the bag that is folded under the flap helps to keep out air and preserve freshness.

A typical horizontal bag-making machine for side weld bags is shown in Figure 5.2.

The film is drawn into the machine, folded around the plow assembly, sealed with vertically positioned bars, and then cut off from the advancing web. This forms the side weld bag described above, with an open mouth that can then be immediately filled with product and sealed along the top, as shown, or picked off, stacked, and shipped to a separate filling location.

If a single web is used on this machine, the depth of the pouch will be limited to 1/2 the film width. By using two webs and making a bottom seal, the pouch depth can be doubled. This latter technique also offers the packager the option of choosing different film constructions for the two sides of the pouch. This is often advantageous in packaging food and other items where a combination of one clear web and one pigmented or metallized web displays the product to best advantage on the store shelf. Alternatively, with the two-web machine, the packager can use only one printed web and a lower cost unprinted web for the other side.

Since they can be operated with a continuous uninterrupted flow of film, these machines can make and fill pouches at speeds up to 1,200 units per minute. This gives them a major advantage over VFFS machines, but this is offset by two drawbacks: they occupy more expensive building space

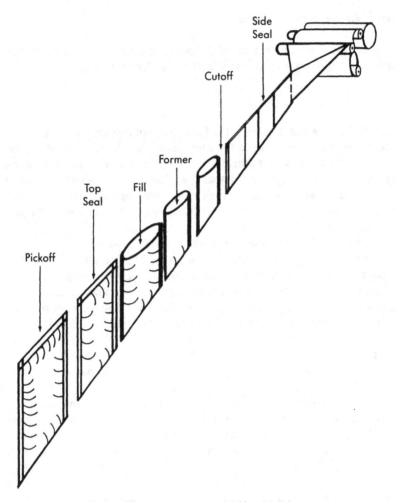

Figure 5.2 A Horizontal Bag-Making Machine.

than do the vertically oriented machines, and films track better on vertically oriented machines. Pouch machines can also be set up vertically (see
below) but only for off-line filling, which is more costly in some cases.

Bottom Seal Bags

These bags are made from tube stock with a transverse heat seal made
to create the bottom. These bags are used when a strong seal is needed,
and are usually made in large sizes for groceries, trash, or leaves. This
method of bag construction has the advantage that it can be set up in line

with the film extruder which creates the tube (see Chapter 2). This avoids the extra step of slitting the extruded tube to make a single layer of flat film. If the extruded tube is larger than the desired bag size, it can be slit with a slit sealer that cuts through the tube in the machine direction and immediately seals the cut edges to form two tubes.

Film Requirements

If product filling is carried out in-line, hot tack is again important, but less so than on a VFFS machine. COF and jaw release are as important here as they are on VFFS machines. If the pouches are made for filling off-line, they must be readily opened with some variant of the air jet device shown in Figure 5.2. This puts a premium on the so-called "anti-block" properties of the film. "Blocking" is the term that denotes the tendency of a film to stick to itself. Coating may be used to diminish blocking for plastics that have this drawback.

LIQUID PRODUCTS

Most liquid products are packaged in rigid or semi-rigid containers made of glass, metal, paperboard or plastic. Some liquid products, however, are packaged in plastic film pouches. Of all the types of containers that are used for liquids, plastic pouches are by far the lowest in cost, since the pouches themselves weigh typically 1/3 to 1/10 as much as rigid or semi-rigid containers. For example, when milk is packaged in a plastic pouch instead of a plastic jug or a paperboard carton, the customer gets one quart convenience at the per-quart cost of a one gallon container

The first challenge presented to the packager of liquids who uses plastic pouches is package integrity. The package wall and the seals must be absolutely leakproof and must be able to survive the mechanical stresses encountered in product distribution.

Frequently the packager supports the pouch by using a paperboard carton as the outer container. The most widespread version of this system is the so-called "bag-in-box" arrangement, which has been used for over 40 years in the U.S., first for battery acid, then for milk, and now for a wide variety of liquid food products. Some liquid food products are packaged and distributed in plastic pouches without a supporting box. The most successful system of this kind is the DuPont Canada pouch system first developed in Europe and now widely used in Canada for milk. Small single serving packages for fluid condiments such as ketchup, salad dressing, and mustard are other examples of the use for liquids of unsupported plastic pouches, but the low product weight in these cases makes the packaging job somewhat simpler.

Unsupported Plastic Pouches

The VFFS system has been successfully adapted for packaging milk and other fluid food products. The machine itself, which must be made of stainless steel and operate in a sterile environment, otherwise runs in exactly the same way as described above for free-flowing dry products. Other than the hygenic requirements, the major difference is in the requirements for the film. Here the top and bottom seals must be made through a film of liquid that frequently contains appreciable amounts of fat. Thus the film must be able to be hermetically and reliably sealed under very difficult conditions which are too demanding for most plastics. LLDPE can be hermetically sealed to itself under these conditions, as can ionomers, but the lower cost LLDPE also has better strength and puncture resistance — properties that pouches holding a quart or more of liquid must possess.

The development of LLDPE film was essential to the commercial success of this system. In addition, the development of a virtually foolproof machine and the immediate availability from the supplier of a cadre of trained mechanics were also necessary, since dairies normally buy containers that are ready for filling and are not accustomed to operating complex package-making machinery. Although the VFFS milk pouch system has yet to penetrate any market outside Canada, its low cost and the ready disposability of its containers may one day make it successful elsewhere.

Small quantities of fluid products such as ketchup, mustard, and mayonnaise served in fast food restaurants are usually packaged in four-side-seal pouches. These packages are made on a vertical machine to which is fed two webs of film. The two webs are brought in contact at their edges and sealed. Several small pouches can be made simultaneously by sealing the two webs together at intervals across their width and then slitting them apart through the middle of each common seal. The bottom seal is then made, product metered in, and the top sealing bar equipped with a cutoff device makes the fourth seal and cuts the packages apart.

This technique has the same advantage as do the other two-web-fed, four-side-seal devices discussed elsewhere in this chapter: the two webs can be made of different materials or the same material printed in two different ways.

Chub Packaging

This rather specialized packaging technique is used for very viscous but nevertheless pumpable products such as certain types of processed meat, cookie and bread dough, processed cheese, and explosives in paste or gel form. As on the more common VFFS equipment, these packaging machines are fed by a roll of plastic and a stream of product. The plastic passes over a forming head where a side seam seal is made to create a tube. Then, instead of a heat seal, the tube is closed with a metal or plastic clip to contain the product which is then metered in. The top of the filled package is subsequently clipped off in the same operation that clips off the bottom of the next tube. This process takes place continuously, so the finished "chubs", so-called, must be cut apart on the fly. This type of clip closure is used when heat sealing cannot be used, as in the explosive application, or when the contents require a specialized film that is not readily heat sealable.

Plastic Pouches Supported by a Box

The use of a paperboard box to support the plastic pouch that contains the fluid product allows packagers to use the bag-in-box system for much larger quantities of product than are feasible in the unsupported pouch system. Individual bag-in-box packages containing up to 330 gallons of liquid products, mainly foods, are now used. These lighter weight packages offer important savings to the packager: a 5 gallon bag-in-box costs about 33 cents per gallon of contents whereas a case of 4.5 gallon steel cans costs about 82 cents per gallon of contents.

The bag-in-box system consists of a plastic bag fitted with a spout that is assembled as an integral part of the bag during the bag-making process. After the bag is filled through the spout with liquid, usually in a separate step, the spout is capped off with a molded plastic valve through which the liquid contents can later be dispensed. The bag is usually a lamination of two different films, one serving as an inner layer that contacts the food contents and the other providing the barrier to gases that is usually required. The bag is made by feeding two webs to a bag-making machine such as the one pictured in Figure 5.2. The machine combines the two webs, one of which has holes punched in it at regular intervals, and seals them together along both edges to make a tube. The spout is inserted into the hole and heat sealed in place. The cross seals are then made to close off the two ends and the finished pouches are cut apart.

The film requirements for this package depend more on the needs of the contents than on the characteristics of the bag-making equipment. EVA, a strong, readily heat sealable, low cost film is the most widely used in single layer bags, but, as noted above, for oxygen sensitive products, additional films incorporating gas barriers such as nylon or metallized coatings are required.

Machinery designed to provide sterile conditions must be used when packaging liquid foods in these containers. The inside of the bag itself is sterilized by the heat of the bag-making process. These conditions are maintained by capping off the bag while it is still hot. When the bag is filled in a separate operation divorced from the bag-making process, its outer surface must be sterilized with steam or chlorine. In cases where film making, bag making, and product filling are all carried out in the same operation, the heat generated in the film-making process is sufficient to sterilize the entire bag.

The various applications now served by the bag-in-box system are described more fully in Chapter 6.

INDIVIDUAL SOLID PRODUCTS

There are many different ways of packaging solid objects in plastic film, and each requires its own specialized type of packaging machinery. Products are frequently packaged in pouches made on horizontal machines similar to that pictured in Figure 5.2. Pouches such as bread bags are made on vertical machines as described below. Other systems include horizontal form-fill-seal, film overwrap, thermoform-fill-seal, shrink packaging, and stretch packaging.

Vertical Pouch Machines

The vertical pouch or bag machine operates in the same way as do the horizontal pouch machines except it is turned vertically (see Figure 5.3). For rapidly perishable food products hot wires are used instead of sealing bars. Although this sealing technique will not produce a reliable hermetic seal, it is a much simpler system from the investment, operating, and maintenance standpoints.

The advantage of vertical operation as opposed to horizontal operation, in addition to the floor space consideration mentioned above, is that it is easier to persuade films, particularly limp films such as LDPE, to track properly on the machine. In effect, gravity helps keep film properly aligned. On the horizontal machines, gravity tends to make limp film drop away from its intended path. The sole disadvantage of vertical operation is that the product must be inserted into the package in a separate operation. Thus it is most often used to make packages for producers such as bakers who have no desire to make their own packages.

Horizontal Form-Fill-Seal Machines

This packaging method is widely used for individual solid items that can be pushed into the filling tube and subsequently into the formed film tube. It is shown in Figure 5.4. The machine feeds the film over a hollow former to create a tube. The product is pushed or conveyed into this tube, which is simultaneously sealed along its longitudinal seam with rotary fin seal

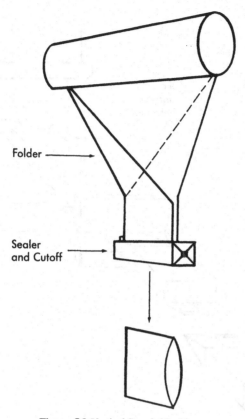

Folder ⟶

Sealer and Cutoff ⟶

Figure 5.3 Vertical Pouch Machine.

wheels. The product and its surrounding tube then move to the end crimp and cutoff station where the final transverse seal is made on one package while the initial transverse seal is made on the following package. A cutoff knife separates the two packages. A crucial element of successful operation is proper integration of product feed with the operation of the machinery. These machines operate continuously at speeds in excess of 400 packages per minute.

The HFFS machine pictured below makes a pillow pouch, but the machine can be adapted to make gusseted bags that are shaped like a rectangular parallelepiped to contain larger objects. These machines may also be equipped with devices to punch holes in the top seam for packages that are hung on racks in the store.

Instead of feeding a HFFS machine with one web that is folded and then sealed, two webs may be fed and sealed on all four sides after product has been added. This technique offers the packager the option of using webs

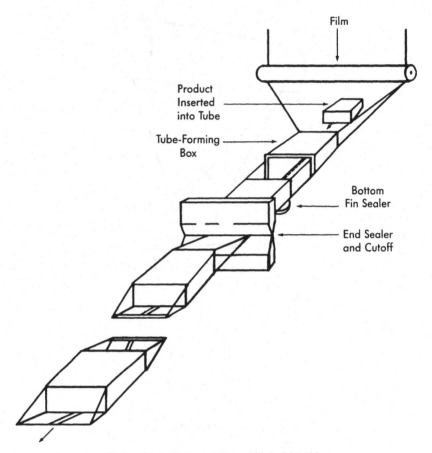

Figure 5.4 A Horizontal Form-Fill-Seal Machine.

made of two different materials or webs made of the same material but printed or decorated differently. For example, the web destined to be the package bottom may be printed opaque. This can enhance the graphics used on the top side which is printed on clear film that displays the product. Many processed meats are packaged in this way, using plastic film for the top web and paperboard for the bottom web.

HFFS machines put more modest demands on film properties: hot tack, for example, is not nearly as important as for VFFS packaging. Adequate slip, modulus, and heat seal capability are about all that are required.

Horizontal Thermoform-Fill-Seal Machines

While these machines closely resemble HFFS machines, they differ in that they require two separate webs and are designed to create a shallow

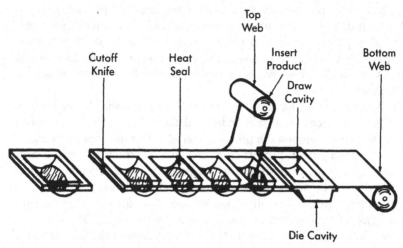

Figure 5.5 A Horizontal Thermoform-Fill-Seal Machine.

tray from the bottom web. This tray holds the product. The top web, which is frequently a different film construction, is designed only to protect the contents. This packaging system is widely used for packaging food items such as fresh red meat. In one variation, the tray is separately thermoformed and fed to the filling machine which inserts product, seals on the lid, and occasionally further overwraps the entire assembly if the tray itself does not have all the properties necessary to protect the contents.

As shown in Figure 5.5, thermoforming of the bottom web is accomplished by pulling the film across a heated cavity, or die, into which it is drawn by vacuum at temperatures high enough so that the film can flow to adopt the shape of the die. This step requires films with special forming characteristics that enable them to survive this process without tearing while hot or without excessive thinning at the edges and corners where the film is distorted to a greater degree.

Some systems of this kind are set up to accept film directly from the film extruder, but an arrangement this complex is rarely popular with packagers, who would rather leave the complications of film making to another manufacturer.

The speed of these machines is limited by the time required in the thermoforming step. This puts a premium on film constructions that can be rapidly thermoformed.

Wrapping Machines

Thus far, the discussion of packaging machines has focused on systems where the package is some form of plastic film: a bag or pouch or lidded tray. Very often, particularly for non-food products, the primary package is a paperboard box. Some boxed products such as cigarettes are over-wrapped in plastic, for further protection or for decorative purposes, on wrapping machines. In some cases, the overwrap is solely for mechanical reasons, as in situations where it is used to bundle a group of objects together for distribution.

The most common form of box wrapper uses non-shrink, non-stretch film fed to the machine either as individual sheets or from a roll, wrapped and tucked neatly around the product, and glued or heat sealed in place. While this describes the process in simple terms, there in fact exist a variety of machines for this purpose which produce many different types of folds and seals. Figure 5.6 shows a few typical examples.

Since these machines usually demand little of the films that are the subject of this book, they will not be discussed in any detail. The reader is referred for further information to the literature cited in the Bibliography. From the film point of view, the main characteristics required are usually those needed to either further protect the product or to enhance its eye appeal. In some cases, however, line speed is crucially important. The most prominent example of this is the overwrapping of the billions of packages of cigarettes that are produced each year. The OPP film that is now universally used for cigarette overwrap is carefully formulated to reduce COF to a minimum and to be successfully heat sealed in dwell times of a fraction of a second. Line speeds in excess of 400 packages per minute are

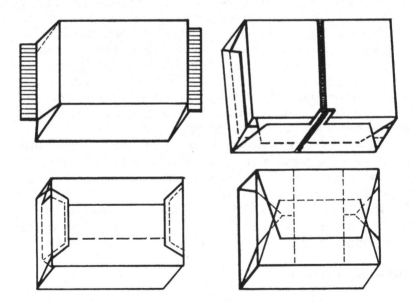

Figure 5.6 Types of Fold and Seal Patterns in Box Overwrap.

routinely achieved on these machines, but only with film that has the right properties.

Shrink Wrapping Machines

Conventional box overwrapping machines were developed when paper was the only flexible wrapping material available. The advent of plastics brought not only clarity but also a variety of special capabilities to the world of packaging. One such capability was the shrink film described in Chapter 2.

The machines that marry these specialty films with their contents all consist of a device to feed the film to a point where it can be wrapped loosely around the product; hot wire sealing devices, which produce a side seam, seal both ends closed and cut off the sealed package; and a so-called "shrink tunnel" through which the loosely wrapped object passes to heat shrink the film tightly around the package.

As with conventional box wrappers, shrink wrapping machines range from inexpensive models used sporadically to wrap objects produced one at a time to more costly machines designed to accept uniformly shaped products that are produced rapidly and that must be wrapped equally rapidly, at speeds up to 150 packages per minute. The simplest and slowest of these machines is the so-called "L-bar sealer", its name derived from the shape of the bar that produces the seals. Figure 5.7 shows how the L-bar sealer works.

Since the L-bar sealer operates with folded film, as shown, the manufacturer must supply his film product in that form.

High speed shrink wrapping machines automate the process, as shown in Figure 5.8, where the shrink tunnel and film supply roll have been omitted for simplicity. To maintain continuous operation, the orbital cross-sealing mechanism must travel along with the film for a dwell time sufficient to produce a strong seal and then return to perform the next seal. The arrows above that device in Figure 5.8 illustrate the necessary reciprocal motion.

Unlike box wrappers, shrink film wrappers demand very special film characteristics. Production of the state of orientation that produces balanced shrink properties is described fully in Chapter 2. A *balanced shrink film* is usually desirable because it produces a more attractive package. Unbalanced shrink properties are advantageous if the item to be wrapped has an odd shape, but the added cost of the rarely used unbalanced film usually persuades the buyer to settle for a balanced film.

Shrink temperature and *shrink temperature range* are also important. The former should be as low as possible and the latter should be as large

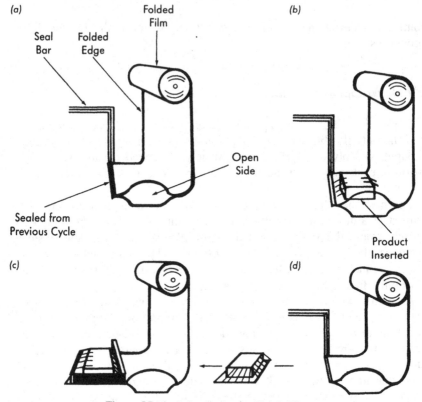

Figure 5.7 The L-bar Sealer for Shrink Film.

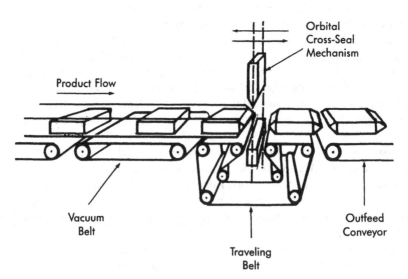

Figure 5.8 An Automatic Shrink Wrapper.

as possible to simplify the operation of the shrink tunnel and to prevent heat damage to package contents.

Shrink tension is crucial when delicate or easily distorted items are being shrink wrapped. For example, most commercially available shrink films exert a shrink force on a thin pad of paper that is so large that the pad tends to curl up into a cylinder after the film wrap is heat shrunk, rather than lying flat.

The *degree of shrinkage* varies from about 25% to about 75% and is shrink temperature dependent. The degree of shrinkage needed depends on the application. Contour wrapping a highly irregular object requires a high degree of shrinkage to produce a neat package; tightening up a loosely wrapped regularly shaped package requires a much lower degree of shrinkage.

For most shrink film applications, sparkling *clarity*, good *slip*, and *clean hot wire sealability* are important. For high speed machines, *stiff films* are greatly preferred by the packager. Since each package must have a small hole punched in the film to allow entrapped air to escape as it shrinks around the package, good *notch tear resistance* is important.

Sleeve Wrap Machinery

The shrink wrap machines described above are used mainly for consumer goods. The shrink wrap concept is also used for industrial tray wrap and bundling where the primary function is assembly rather than enhancement of consumer appeal. Sleeve wrap bundling is used for many kinds of products such as books, office supplies, ice cream cartons, newspapers, and boxes of canned goods.

Sleeve wrappers, pictured in Figure 5.9, are fed by two rolls of film. After they have been sealed together along two edges, product is conveyed into the sleeve of film thus created. Passage through a shrink tunnel shrinks the film tightly around the contents but leaves holes in both ends. While this is a drawback if total wrapping is important, it does provide convenient hand-holds for carrying and eliminates the need for a hole in the film to allow the escape of air during the shrink step.

Film designed for sleeve wrapping must possess the same shrink characteristics as film for shrink wrapping, but is usually thicker to give it the necessary strength to handle heavier loads. Since esthetics are far less important in this industrial application, film optics, crucially important for consumer goods shrink wrapping, are unimportant, so lower cost films emphasizing strength and puncture resistance rather than excellent optics are normally used.

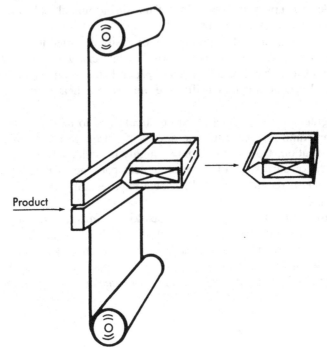

Figure 5.9 A Sleeve Wrapper.

Stretch Wrapping

Stretch wrapping is a process that assembles a load. It is used for unitizing large loads on pallets for shipment. The traditional method of unitizing pallet loads used straps made of metal or plastic. Stretch wrapping in plastic offers securer unitizing, sometimes at lower cost, than these traditional methods, since the plastic wrap surrounds the entire load and can stretch and sway, much like a rubber band, under the influence of shifting stresses on the load. In the U.S. today, well over 250 million unitized loads are shipped with stretch wrap protection, and stretch wrap film has become the largest single market (about 40 million lbs/yr) for film grade EVA resins.

In conventional stretch wrapping, the product load is placed on a rotating turntable. As it rotates, the load pulls the film from the roll, as shown in Figure 5.10. The film stretching is accomplished by restricting the film unwind movement with some sort of braking device on the film roll. Typically, the film is stretched from 10 to 50% as it is applied to the load.

The cost of this packaging method can be reduced by 20–30% by pre-stretching the film. This is done as shown in Figure 5.11 by passing the film through two rollers, the second of which runs faster than the first, before it is wrapped around the load. Since this is a far more controlled stretching process, the film can typically be stretched about 250%, which substantially reduces the film cost per load. In addition, this degree of stretch orients the film and strengthens it.

The resins commonly used to make film for stretch wrapping are EVA and LLDPE. The important film characteristics are stretchability, stretch force, restretch force, and breaking strength.

Stretchability embraces several more fundamental film properties: elasticity, tear resistance, and puncture resistance. Ultimately, high stretchability means less film per pallet and lower wrapping costs.

Stretch force is a measure of the static force exerted on the load by the stretched film. Very high stretch forces will distort the load.

Restretch force (really elasticity) is a measure of the "rubber band" effect that allows the film to adapt to load sway during shipping without breaking.

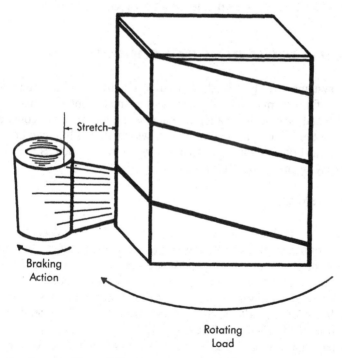

Figure 5.10 A Conventional Stretch Wrapper.

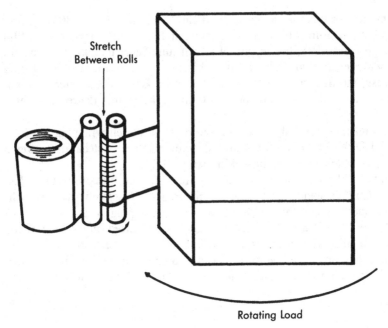

Figure 5.11 A Pre-Stretch Wrapper.

PACKAGING LOOSE ITEMS IN A CONTAINER

Several systems already described are used for this purpose. One is the thermoform-fill-seal machine which creates a small container from one web of film and heat seals another web across the top after the container is filled. Another is the plastic film pouch, which after filling can be heat sealed or closed on top with a mechanical type closure such as the "Zip-Loc" for easy opening and reclosing.

Blister Packaging

A third system is the so-called "blister package". This consists of a stiff, clear, thermoformed plastic box containing the loose product but lidded with a sheet of plastic or coated paperboard that carries the graphics and is sometimes perforated for rack-hung display. Blister packages are used not only for small hardware items but also for luncheon meats, cheeses, hot dogs, health care products, cookies and snack foods. They lend themselves well to gas-flushed, sterilized, or vacuum packaging.

Blister packages are produced on an assortment of machines of varying complexity. In the simplest, blisters are made separately and then loaded by hand with product and lidded by the packager. More complex machines form the blister, load product, seal on the lid, and cut the finished packages apart continuously and automatically. A diagram of blister packages is shown in Figure 5.12.

In operation, the plastic container, either preformed or made directly on the packaging machine, is filled with product and then sealed to the plastic or adhesive-coated paperboard lid by either of the two methods shown in Figure 5.12. Heating time and temperature are both controlled to match the film and adhesive characteristics to provide a strong seal.

Skin Packaging

A fourth technique involves the so-called "skin package" which, like the blister package, normally uses printed paperboard as one side of the pack-

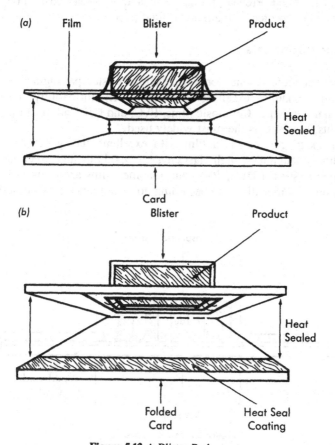

Figure 5.12 A Blister Package.

age. Unlike the blister package, the skin package has a very clear, tough plastic film that is drawn down by vacuum around the contents until the film conforms so faithfully to the product contours that it becomes like a skin and is virtually invisible. This skin film holds the product tightly in place on the printed paperboard, provides excellent tamper evidence, protects it from dust and dirt and preserves the legibility of the printing on the paperboard. Figure 5.13 shows a skin package.

In skin packaging, the product is positioned on the printed, adhesive-coated card which is then moved onto a platen that contains air passages connected to a vacuum system. Plastic film is held in a frame above the product/paperboard card, where it is heated for softening and then dropped, in the frame, onto the carded product. Vacuum is applied to bring the film into intimate contact with the product and card. Residual heat in the plastic film activates the heat seal coating on the card, creating a fiber-tearing bond. Preheat, timing cycle, and post-heat must all be carefully controlled to create an attractive final package.

Film Requirements

The salient requirements for a satisfactory blister packaging film are clarity, high modulus, excellent deep draw thermoformability and good sealing characteristics. Cellulosic films, PVC, and HIPS all meet these requirements, but PVC is the most widely used.

Skin packaging demands a film with excellent clarity coupled with extraordinary toughness, high elasticity, low modulus and excellent heat seal characteristics. LDPE, PVC, and ionomer films are all used in this process, but ionomer films are becoming dominant since they possess the

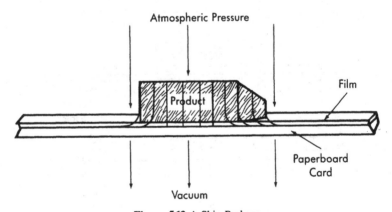

Figure 5.13 A Skin Package.

best balance of the critical properties that are needed. Their higher cost per pound is offset by their higher strength, which allows downgauging to produce a competitive cost on an area basis.

HEAT SEALING EQUIPMENT

This section covers the various heat sealing devices in common use on packaging machinery designed to handle plastic films. Although some of these devices tend to be specifically oriented towards one machine type rather than another, it is convenient to discuss them all here so the reader can clearly see their differences and similarities.

All these devices are designed to bring two separate layers of film together, heat them up until a seal is made, and then separate, allowing continued film travel. Some heating jaws also incorporate a sharp blade that cuts through the seal to separate the finished package.

Impulse Sealing

Impulse sealers consist of lightweight jaws to which a Teflon® covered resistance wire is affixed. A brief pulse of electric current rapidly heats the wire to a temperature sufficient to melt the plastic in the seal area. After the current is switched off, the jaws hold the seal together until it is cool enough to release. This sealing technique is limited in the width of the seal that can be made: wide seal widths take too long to cool down.

Heated Jaw Sealing

In this method, the jaw remains hot and the time of contact between jaw and film is adjusted to produce a satisfactory seal. These devices work more rapidly than impulse sealers because they usually include cool jaws in tandem to allow the seal to be held together without delaying the further use of the heated jaws. The jaw can be built to operate in a rotary or a reciprocal fashion.

Hot Air Sealing

This method employs a blast of hot air to provide the necessary heat. It is often employed for heavy duty bags containing product such as fertilizer because it employs no mating surfaces that may become contaminated with the product being packaged. Dust particles that become trapped in the seal area do not affect its hermetic property because they become completely surrounded by molten plastic.

Hot Wire Sealing

These devices supply heat by means of an unsupported hot wire that also serves to cut the film. This sealing method is used on shrink wrapped packages because it makes an unobtrusive seal that is esthetically desirable and because its inherently lower seal strength, which results from the very narrow seal width, is not a drawback on shrink wrapped packages that do not have to hold the weight of product dropping into the package.

Ultrasonic Sealing

In this sealing method, an ultrasonic head, called a horn, transmits vibratory energy through two pieces of plastic film held in contact. At the film interface, the vibratory energy is converted to heat, producing an almost instantaneous weld. Conversion of vibrational energy to heat at the interface occurs because at the frequencies employed, the two film surfaces hammer against one another, generating heat by friction.

Normally only similar plastics can be joined by this technique, but this is not a major drawback. It is particularly useful for rigid, low melting plastics or for oriented film, since only the interface is heated. It is also useful when seals must be made between surfaces contaminated by fats and oils, or in cases where thick layers of paper are being sealed together via thin plastic coatings on their surfaces.

The main disadvantage of this method is that it tends to be much slower than the radiant heat methods described above.

Problem Seals

Oriented Film

Oriented film is more difficult to seal because when warm it tends to revert to its unoriented state. This leads to puckering in the seal area. This problem can be overcome by using multi-point sealers which have raised hot spots on the sealing jaws rather than a flat surface. This limits the puckering to isolated spots along the seal, but cannot be used for a hermetic seal. A more common approach is to coat the oriented film with a lower melting polymer that effects the seal at temperatures below the puckering range of the oriented film.

Narrow Sealing Range Films

Films with a narrow heat sealing range are difficult to seal because of the very close temperature control required. Copolymerization to widen the range can be used, but another approach is to use dielectric heating. High frequency electric current is passed through the film by the sealing bars. If the film contains polar molecules, as do PVC and nylon for example, these molecules oscillate under the influence of the current, and this molecular agitation is converted to heat sufficient to melt the film in the seal area. Since this heat is produced uniformly across the width of the film, there is no risk of the outside surface that contacts the sealing bars being overheated in order to bring the inside surface up to the sealing temperature. This method also works well for thick films, for the same reasons. Unfortunately, it will not work for non-polar films such as LDPE, HDPE or LLDPE.

SUMMARY

As noted in the introduction to this chapter, one of the main purposes of this discussion of packaging machinery is to highlight the demands that the machine puts on the film and how film manufacturers must react to those demands. This chapter material is best summarized, therefore, by returning to this theme in a more general fashion than has been possible in the discussion of individual machines.

Both packagers and machinery manufacturers are constantly striving to increase the productivity of packaging machinery. From the film point of view, this means they want a film that will run easily and rapidly through the machine and that can be readily heat sealed to produce seals that have enough mechanical reliability and enough gas impermeability to satisfy the needs of the product being packaged. The term "enough" is used here to emphasize the point that not all seals need be hermetic and that seals on packages that hold an ounce or two of product need not be as strong as those for packages that contain several pounds of material.

The film manufacturer must translate these requirements into film properties that can be measured, thus enabling him to judge whether a new film will be runnable or whether his standard film production continues to meet machine requirements.

The film properties that are germane to these runnability requirements are stiffness, COF, sheet flatness, roll formation, thickness uniformity (gauge control), and a series of properties that govern heat sealability:

- seal temperature range
- jaw release
- degree of hot tack
- percent shrinkage in the heat seal temperature range
- seal strength
- melting point
- melt viscosity

As noted earlier, low COF means the film will move rapidly and easily over all the stationary components of the machine with which it comes in contact. Good flatness, roll formation, thickness uniformity and stiffness mean that the film will proceed uniformly through the machine instead of wandering back and forth and forcing the operator to slow down production to make frequent alignment adjustments.

A wide seal temperature range of, say, 50°F or more is important because it is difficult to control sealing bar temperatures very precisely. If the seal temperature range is less than 10°F, some packages will be incompletely sealed and some will be burned through in the seal area as sealing bar temperatures oscillate around the optimum temperature setting. These packages will be either rejected in later quality control inspection, which reduces productivity, or escape undetected until the customer sees them, which is worse.

Even though sealing bars are frequently coated with low-stick materials such as Teflon©, a polymer that has poor jaw release and sticks tenaciously to hot surfaces will be difficult to handle, finished packages will tend to hang onto the bars rather than dropping free, and occasionally bad seals will result, all of which adversely affect productivity.

Hot tack was discussed in some detail above in connection with VFFS machines. As noted there, this property is particularly important for high machine speed and productivity when package loads are heavy.

Extensive puckering of the film in the seal area will result if the film has greater than 1% shrinkage in the heat seal temperature range. This film drawback can usually be offset by operating at the low end of the seal temperature range, but that requires longer seal dwell times to produce strong seals; long dwell times mean slower machine operation.

Films with high melting points and/or high melt viscosity are disadvantageous in that it takes longer to bring the film temperature up to the sealing temperature range and a longer dwell time at that temperature is required to produce enough resin flow to make a good seal.

High inherent seal strength means:

- fewer packages are rejected as inadequately sealed
- sealing bar temperatures need not be precisely controlled, which avoids frequent shutdowns to make bar temperature adjustments
- the thickness of the seal coating, if any, on the film can be reduced and material costs are therefore lower
- thicker packages can be run on the machine since less heat need reach the seal area to produce the minimum adequate seal integrity; less heat means less distortion, shorter dwell times and thus faster operation.

Like the slip and tracking characteristics discussed above these relevant characteristics can be measured quantitatively. Melting point, melt viscosity, seal temperature range and percent shrink at seal temperature are all straightforward. Seal strength is best judged by a peel test. Hot tack is assessed by measuring the temperature range over which a newly formed seal resists rupture under a given peel force. Jaw release is determined by measuring the force required to pull a newly formed seal away from the sealing jaws. By these measurements, the film producer can characterize the machinability of a new film candidate and can adjust its properties where needed before running it on a machine, even though only actual machine trials will show with certainty how well the film will perform.

Several formulating techniques are available to enable the film producer to correct machinability defects in an otherwise desirable film. Many of these are discussed in Chapter 3. Acrylic coatings or additives such as finely divided silica can be used to lower COF to acceptable levels. Heat seal properties can be improved by coating a film with poor seal properties with a thin layer of plastic resin that has good heat seal properties. OPP and other oriented films that are all difficult to seal without puckering and distortion are frequently coated in this way. For example, a PP/3–5%PE copolymer can be coextruded with PP and the coextrusion then oriented to produce OPP with an unoriented seal layer of PP/PE. Alternatively, OPP can be coated with PVDC to produce a seal layer that also improves the barrier properties of the base film.

BIBLIOGRAPHY

Bakker, M., ed. 1986. The *Wiley Encyclopedia of Packaging Technology.* New York, NY: John Wiley and Sons, Inc..

Briston, J. H. 1989. *Plastics Films.* Harlow, England: Longman Scientific and Technical, 3rd edition.

[1]Davis. 1982. *Packaging Machinery Operations – Bag Making, Filling, and Sealing.* PMMI.

[1]Davis. 1982. *Packaging Machinery Operations – Blister Packaging and Thermoform-Fill-Sealing.* PMMI.

[1]Davis. 1982. *Packaging Machinery Operations – Form-Fill-Sealing.* PMMI.

[1]Davis. 1982. *Packaging Machinery Operations – Wrapping, Overwrapping, and Bundling.* PMMI.

Hanlon, J. F.. 1984. *Handbook of Package Engineering.* New York: McGraw Hill.

Jenkins and Harrington. 1991. *Packaging Foods with Plastics.* Lancaster, PA: Technomic Publishing Co.

[1]These references contain detailed descriptions of the configuration and operation of all types of packaging machinery in the categories listed in the title. Each step is clearly described, with the description supplemented by a wealth of drawings of the various machine components. Filling equipment and the other necessary adjuncts to the packaging operation are also described in detail. These publications are available from PMMI, 1343 L St. N.W., Washington, D.C. 20005.

CHAPTER 6
Packaging Applications for Plastic Films

INTRODUCTION

This chapter covers the major packaging applications for plastic films in the U.S. The contents are divided into sections, each dealing with a major application or market. For each market, we discuss the films that are used and why those films are most appropriate for that application. This requires repeated reference to film properties. To make the presentation more readable, these properties are discussed in qualitative terms. Reference to Table A.1 will provide the necessary quantitative basis for the qualitative discussion.

Statistical data gathered from many sources is included to show the relative importance of each film type in each market. All data apply to the year 1989 unless otherwise indicated.

Some large markets, such as food, are important users of multilayer films; others, such as packaged hardware or merchandise bags, use mainly monolayer films. Chapter 3 covers the production of multilayer films. This chapter has been written assuming that the reader has mastered that material, so the terms "lamination" or "coextrusion" will be the only description of the process by which multilayer films have been manufactured for the application under discussion. The usual convention for identifying the structure of multilayer films is used: the outside layer is identified first and then each successive layer through to the inside layer, e.g.:

OPP/tie layer/EVOH/tie layer/PE

In this structure, OPP is the outermost layer that contacts the environment and PE is the innermost that contacts the package contents. As this example also shows, abbreviations such as OPP and PE will be used for the various plastics involved in these applications. Reference may be made to Chapters 1 or 2 or to the Glossary in the Appendix for clarification of these abbreviations.

The frequent use of multilayer films greatly complicates the assembly of

179

statistical data. A multilayer film that contains, for example, three separate different film layers is included in the data as if it were three separate films. Although this approach gives the reader the best feel for the quantities of the various plastics that are used, it also requires many approximations to be made in the data themselves. Thus it is possible that any given figure in the tables may be in error by 10–20%. While greater accuracy would certainly be desirable, it simply is not possible, and will not be until the day when better data are available.

Table 6.1 shows the total 1989 U.S. consumption in packaging applications of the major plastic films.[1]

These data clearly show the importance of the packaging market and food packaging in particular as a consumer of plastic film: packaging consumes 90% of the total film and food packaging consumes about 35% of the total packaging film. Merchandise bags and trash bags run food packaging a close second at 26%, so those two applications account for more than 50% of the packaging film consumption.

PET is the only film included for which flexible packaging is a minor market. About 20% of the total PET film market is packaging. The major film applications for PET are electrical, magnetic tape, and imaging.

Growth of plastic film volume in 1989 vs. 1988 was generally in the 2–4% range for most films except LDPE and PVC. LDPE volume declined due to its replacement by higher strength LLDPE and HDPE. PVC declined due to the drop in consumption of fresh beef, a large packaging application for PVC film.

More detailed data on the individual packaging applications for LDPE and HDPE are shown in Table 6.2.[2]

Unlike the data in Table 6.1, these LDPE data include LLDPE and EVA, but separate EVA data are shown in Table 6.1. Since the data sources and collection methods were not identical, minor differences will be found in the various subcategories. For our purposes the differences are not important. Although data for LDPE, EVA, and LLDPE are all combined as LDPE in Table 6.2, some of the application categories are clearly dominated by *one* of these three different films. For example, dairy is mostly LLDPE, rack and counter bags are mostly LDPE, and so forth.

A breakdown of the food packaging market for PVC film is shown in Table 6.3.[3] The large institutional market consists of film used in institu-

[1]1990. *Modern Plastics*, "Monthly Statistical Report—Resins," SPI Committee on Resin Statistics as compiled by Ernst & Young (January):106.

[2]Ibid. p. 100.

[3]Bakker. 1986. *The Wiley Encyclopedia of Packaging Technology*, New York: John Wiley & Sons, p. 309.

Table 6.1. 1989 consumption of plastic films in packaging.

Film	Consumption (Million Pounds)
HDPE	830
Food	144
Non-Food	70
Merchandise Bags	306
Trash and can liners	139
Other Packaging	107
Other Non-Packaging	64
LDPE	3443
Food	1132
Non-Food	989
Shrink & Stretch Films	247
Merchandise Bags	105
Trash and can liners	372
Other Non-Packaging	598
LLDPE	2402
Food	197
Non-Food	618
Shrink & Stretch Films	282
Merchandise Bags	198
Trash & can liners	894
Other Non-Packaging	213
PP	541
Oriented (OPP)	
Snacks	100
Baked Goods	60
Candy	30
Tobacco	30
Cheese, Coffee, Tea	19
Labels	15
Box Overwrap	30
Pasta, nuts, misc. food	25
Non-Packaging	100
Unoriented PP	132

(continued)

Table 6.1. (continued).

Film	Consumption (Million Pounds)
PS	211*
PVC	367
Packaging	320
Non-Packaging	47
EVA	640*
7% or above VA content	
Pallet stretch wrap	15
Ice bags	30
Meat barrier bags	40
coex sealant layer	25
bag-in-box	20
other specialties	10
Below 7% VA (see LDPE)*	500
Nylon	58*
PET (est.)	75
Total Packaging	7545
Total Film	8567

* In the table, the data for PS and nylon are total film consumed. It is a safe assumption that over 90% of the film in these categories is used for packaging. The EVA data are divided into two categories depending on VA content. EVA film at less than 7% VA closely resembles LDPE, is produced by many manufacturers who also make LDPE, and is used as a substitute for LDPE in many LDPE applications, principally industrial, trash and garbage, and merchandise bags. Absent more accurate figures, it is safe to assume that 90% of this 500 million pounds of EVA goes to those applications.

tional kitchens and restaurants and by caterers to overwrap food plates, trays, sandwiches, glassware and utensils.

As noted in Chapter 5, shrink wrapping in polyolefin or PVC films is used in certain packaging applications to enhance package appearance, bundle several items together, provide tamper evidence and serve as a dust cover. There are two separate shrink film markets. The *consumer market* demands high clarity so the film wrap will not obscure the contents and will add to product appeal. The *industrial market* uses lower cost, thicker gauge shrink film to bundle several items together. These industrial applications can be very large, both in terms of film consumed and in terms of the size of the package being wrapped. For example, entire pallet loads weighing many tons are sometimes shrink wrapped to hold them together for shipping.

A breakdown of the *consumer market* for high clarity polyolefin and PVC shrink films is shown in Table 6.4. Growth in most of these markets is a modest 2–3%, but food applications are growing at over 10%.

FOOD PACKAGING

Americans spend about $500 billion each year to feed themselves. Almost all the food they consume is packaged in some way at one or more

Table 6.2. Packaging markets for HDPE and LDPE film.

HDPE

Market	Consumption (Million Pounds)
Merchandise Bags	182
Tee Shirt Sacks	215
Trash Bags	
Institutional	124
Consumer	15
Food bags and box liners	110
Multiwall sack liners	50
Other Packaging	70
Total HDPE in Packaging	766

LDPE

Food Packaging	
Baked Goods	310
Candy	50
Dairy	72
Frozen Food	108
Meat/Poultry/Seafood	205
Produce	158
Non-Food Packaging	
Tee Shirt Sacks	159
Other Merchandise Bags	125
Grocery Wetpack	91
Self-service bags	99
Garment Bags	134
Heavy Duty Sacks	156
Industrial Liners	203
Rack and Counter Bags	228
Multiwall Sack Liners	47
Pallet Shrink Wrap	45
Shrink Wrap other	170
Stretch Wrap	260
Textile	196
Trash Bags	1310
Misc. Food, Non-Food	605
Total LDPE in packaging	4731

Table 6.3. Distribution of PVC film among food packaging applications.

Application	% Film Consumed
Meat	57
Poultry	4
Produce	9
Fish	4
Frozen food	1
Institutional	24
Baked Goods	1

Table 6.4. The high clarity shrink film market.

Application	Volume (Pounds in Millions)
Printing, Paper, Stationery	30
Toys and Games	9
Foods	24
Hardware/Housewares	10
Records and Tapes	3
Pharmaceuticals	2
Other	12
Total	90

points in the food distribution chain. Food dominates packaging in another way: of all the products packaged in the U.S. today, about half are food products. From the film point of view, this application consumes over 90% of all the OPP, PP, PVC, nylon, and PS films that are produced, about 15% of the HDPE film, and about 25% of the LDPE film, the latter category including LLDPE and EVA. It also involves a greater variety and complexity of film structures than any other packaging film market.

All foods are perishable, and as they age, they become less palatable, lose their nutritional value and sometimes become dangerous to consume. Many food products are thermally processed before they are packaged to destroy bacteria. Refrigeration slows bacterial growth and is used for foods such as milk that are not fully sterilized by heat. Yeasts and molds are generally found in association with high acid sugary products and grow best on foods exposed to air. Even in the absence of air, low acid foods will support the growth of the *Clostridia* organisms that are capable of producing botulism toxins.

Maintenance of the correct water content is important for extended shelf life. Dry foods, if they become moist, are subject to increased biochemical degradation. Water loss from wet foods will alter physical characteristics and can lead to increased microbiological growth. All these considerations mean that foods must be shielded from harmful environmental factors by some kind of protective package.

Food processors have a responsibility that in degree is shared by few other consumer goods manufacturers – the continuing good health of their customers. At the same time, they are in business for profit, which requires them to continuously seek the lowest cost package that meets all their needs.

To further complicate matters, foods are produced and consumed in seemingly infinite variety, and thus differ widely in their need for protection against an assortment of hostile environmental factors. All the reasons given above – perishability, health, product variety, and cost – account for the tremendous variety of package types and film structures used and the billions of dollars spent on materials for this packaging application.

Fresh Foods

About $45 billion worth of food products in the U.S. consists of animal meats: beef, poultry, fish, mutton, veal, and pork. Fresh produce accounts for another $30 billion. These products are subject to microbiological and physical changes that require sophisticated and relatively expensive packaging to extend shelf life past a few days. Plastic films have become the dominant packaging material for almost all of these products, since these films provide the best combination of protection, product display and low cost.

Beef

The color of red beef depends upon the presence of oxygen. The natural color of myoglobin meat pigment is purple. When exposed to air, this pigment is converted to oxymyoglobin, which is red. When oxymyoglobin is further oxidized, it turns brown. Understanding this color sequence is crucial to understanding why fresh beef is packaged as it is.

Beef is packaged in plastic films for two separate and distinct purposes: distribution and selling at retail. The packaging films used are quite different in these two situations.

For distribution from the slaughterhouse to the supermarket, beef is packaged in hermetically sealed heat-shrinkable bags made from a three-layer coextrusion: EVA/PVDC/EVA. In this structure, EVA provides the toughness and hermetic seal characteristics while PVDC provides the high oxygen and moisture barriers that are needed to prevent premature oxidation and moisture loss. This coextrusion is made as a shrink film so that after being filled with beef, it can be evacuated and shrunk around the product in hot water to eliminate residual oxygen that would otherwise remain in the package.

When the beef reaches the supermarket, it is unpacked, cut by the store butcher into consumer-size portions, and repackaged on a plastic or paperboard tray overwrapped with PVC film. PVC has excellent clarity, gloss, and toughness combined with the right oxygen permeability and moisture impermeability to prevent moisture loss while allowing access to oxygen which converts the purple myoglobin meat pigment to red oxymyoglobin. Customers associate the red color with freshness, so shelf life, which would be greater were the beef packaged in an oxygen barrier film, is sacrificed to satisfy this consumer mindset.

The low cost of PVC and the excellent match of its properties to the requirements of the product make retail beef and poultry packaging the largest food packaging application for this plastic film. Similarly, the EVA-based barrier bag used for distribution is the largest single application for EVA in food packaging.

Rival systems for both distribution and retail packaging nevertheless exist. Those that avoid the use of PVC are particularly popular in Europe, where concern about the supposed harmful environmental effects resulting from incineration of PVC is an important factor in the producer's selection of packaging material.

Of far greater importance to the future position in the U.S. of PVC in retail beef packaging is the trend to "case-ready" products. These are retail beef cuts created in the slaughterhouse and shipped to the supermarket ready for display and sale. Large beef producers greatly prefer this marketing approach because it allows them to brand their products on the store shelf and gives them both better control of quality and a greater share of the total profit from the sale. This trend, although inevitable, is not a rapid one. In-store butchers don't like it and it demands a more complex package that can both preserve the beef during distribution and allow it to be sold with the red color consumers prefer. While they experiment with such packages, producers are also trying to persuade consumers that purple meat is actually fresher. The latter approach may be more successful in the long run and will certainly be less expensive.

Pork

Pork is packaged in plastic films in much the same way and with the same films that are used for beef. Since the red color is not a major consumer issue, processors can use preservatives combined with vacuum or modified atmosphere packaging (MAP) to facilitate the adoption of the case-ready approach. One such system packages retail cuts of pork on EPS trays overwrapped with PE or PP/PE shrink film. These small packages are then combined in a gas-purged master pack consisting of a PET tray lidded with a PET/PE/ionomer film. The whole assembly provides a 14 day shelf life, long enough for safe distribution. When the individual units are removed, they have about a 3 day shelf life in the supermarket.[4]

Poultry

As with pork, the red color issue does not exist for poultry. This simplifies the task of the package designer and leads to the more frequent use of polyolefin films as an alternate to PVC in fresh poultry packaging.

[4]1987. *Packaging Digest* (July):32.

Today's standard poultry package consists of an EPS tray overwrapped with a stretch PVC film or a polyethylene or polypropylene copolymer shrink film that can first be stretched over the package and then heat-shrunk to hold the packaged parts firmly in place. Since the stretch PVC packages occasionally tend to leak, the firmer wrap obtainable with PE or PP copolymer shrink films is gaining in popularity.

Poultry producers are capitalizing on the increasing consumer preference for their product as opposed to beef by offering many value added products that are marinated, breaded, or contain spices to simplify preparation of tasty meals in the home. These newer products require more complex packages since they tend to move more slowly off the supermarket shelves and thus must have longer shelf life. To achieve this goal, most producers now use MAP in multilayer trays fitted with multilayer film lids. The lids are usually laminations that incorporate PET or nylon for physical protection, PVDC or EVOH for barrier, and EVA or ionomers to produce hermetic seals. PET film bags are sometimes used for processed poultry parts such as turkey nuggets and sticks. The film is metallized for increased protection against oxygen.

In summary, PVC film still dominates the poultry market. Polyolefin shrink and multilayer barrier films, while growing rapidly, have only a minor share.

Fish

Because fish products are highly sensitive to enzymatic spoilage and the development of toxic bacteria, they are either sold fresh with a very short shelf life, packaged in modified atmospheres, or frozen and then packaged. Of these three alternatives, most fish are sold frozen with plastic films such as OPP, PET, or HDPE used to overwrap the product inside its paperboard box that also may be overwrapped. The primary function of these films is to prevent moisture loss. Fish high in fat content requires an oxygen barrier such as PVDC added to the base film.

Most fresh fish is sold unpackaged, but a modified atmosphere packaging system has been developed that is safe but rather complex and thus rarely used. The film involved must possess high oxygen permeability; LDPE has this essential characteristic.

Produce

Processor packaging of fresh fruits and vegetables in plastic films is still in its infancy. An explanation for this requires some background. A food processor who wishes to package fresh produce to increase its shelf life must choose a plastic film that has the correct permeability to oxygen, water vapor, and CO_2 and that will allow the maintenance of a modified atmosphere within the package consistent with the need for the product to respire, produce ethylene, and ripen. The optimum atmosphere for each product is different, ranging from 10% O_2, 7% CO_2 for peas to 1% O_2, 10% CO_2 for avocados. Thus the best film for peas will not be the best for avocados. Of all the commerical films available, PVC and LDPE have the best O_2/CO_2 permeability ratio for fresh produce and are used in today's relatively few CAP produce packaging applications. For CAP to be widely adopted for produce, special CAP films need to be developed. This involves, at least initially, high specialty film costs and an expensive development process for each product.

Since most produce processors are not willing to undertake this task in order to extend the shelf life of their products, the major plastic film used for produce packaging today is LDPE. It serves principally to keep the product clean, reduce moisture loss, and provide a convenient way of bundling several items together for store carry-out. Some produce film is perforated to facilitate respiration. As shown in Table 6.2, this is the third largest food packaging market for LDPE, but the *potential* market is far larger for more complex plastic films that would significantly extend produce shelf life and greatly reduce spoilage.

PVC, the next most popular plastic film for fresh produce, is used mainly for overwrapping trayed products where its high clarity and gloss enhance package appearance.

Shrink film packaging systems have been developed for fresh fruit, based on PE or PE-PP copolymer films. This system does an excellent job of preserving fruit by retarding moisture loss, but is generally regarded as not cost effective by most fruit packers.

Frozen produce appears on the supermarket shelves either in paperboard boxes waterproofed with a layer of PE or in plastic bags. These bags were originally LDPE or HDPE, the latter used when greater strength was needed. Now packagers are increasingly turning to coextrusions. For example, EVA/white LLDPE/EVA provides the excellent seal properties of EVA coupled with the good strength properties of LLDPE which is buried to prevent scuffing of the white ink. Coextrusions such as these can be offered at lower cost because overall film thickness can be reduced by about 10%. Another example of this trend is an HDPE/EVA coextrusion where HDPE furnishes not only the required strength but also good moisture barrier; EVA again provides the required seal characteristics. In this application, EVA is often blended with other polyolefins to widen its seal temperature range and furnish the easy-open feature increasingly demanded by consumers.

Some frozen vegetables and other foods are packaged in plastic film pouches so they can be cooked therein by heating in a pot of boiling water, avoiding later cooking pot cleanup. Since these bags must withstand boiling water, PET film, with its excellent high temperature properties, is usually used. A PE layer provides the necessary hermetic seal, since PET, always oriented for packaging applications, cannot be heat sealed to itself without a degree of distortion that makes hermetic sealing next to impossible.

Processed Meat and Poultry

These are products such as frankfurters, sausages, luncheon meat, bacon, dried beef, and processed turkey and chicken parts. Unlike fresh meat, they are always packaged by the food processor and thus need additional protection against oxygen uptake and moisture loss to furnish the long shelf life they must have due to lengthy distribution chains and the irregular purchasing patterns of consumers in the store. Unlike fresh meat, the red color of processed beef products is maintained by adding special ingredients in the processing plant, where salt, spices, and curing agents such as sodium nitrite are also added. To satisfy the long shelf life requirement, multilayer films that incorporate good oxygen barrier films such as PVDC, EVOH, or 6 nylon are the usual choices of the packager.

To accompany the oxygen barrier, seal layers that provide an excellent hermetic seal are used. This makes this application one of the largest markets for ionomer and acid copolymer films and resins. Their outstanding seal integrity and tolerance for grease contamination has led to their widespread displacement of LDPE and later EVA as the seal resin. More recently, the higher cost of these specialty ethylene copolymers has led packagers to use blends of ionomers or acid copolymers with lower cost EVA.

Many of these products are packaged in thermoform-fill-seal equipment that requires good draw properties for the forming web. 6 nylon satisfies this requirement and the oxygen barrier requirement. The property combination possessed by 6 nylon, which also has excellent toughness, makes fresh and processed meat applications the largest food packaging market for 6 nylon film in the U.S. In cases where nylon's oxygen barrier is inadequate, a layer of PVDC or EVOH is added. Either nylon or PET is used as the main lidding component. Top and bottom webs are heat sealed with polyolefin seal resins: LDPE, EVA, or ionomer/ acid copolymer depending on the seal requirements.

Thermoform-fill-seal packages for luncheon meats involve a more rigid container than do link products. HIPS or PVC, both stiff films, are used in this case with oxygen barrier supplied by a layer of PVDC. The lid is a PVC/PVDC coextrusion. The good seal properties of PVDC allow it to perform double duty in this package.

Whole or half hams are usually packaged in shrink bags using the system described above for subprimal cuts of fresh red meat. For hams designed to be cooked in the package, composite nylon/EVOH/ionomer films are used.

Processed meat varieties that can be extruded, such as sausage, liverwurst, and sandwich spread, are chub-packaged on special form-fill-seal equipment in tightly filled film cylinders, or chubs, closed at each end with a twist of wire rather than a heat seal. PVDC is used when good oxygen barrier is needed, or LLDPE in the few cases where its poorer barrier will suffice.

Milk

The very large milk business in the U.S. consumes almost no flexible plastic film. In Canada, about 50% of the fresh milk is packaged in LLDPE film pouches on form-fill-seal equipment, making this the largest food packaging application for LLDPE film in North America. The LLDPE used, a butene copolymer, has the right combination of toughness, seal integrity when heat sealed through a layer of milk, and low cost to effectively compete with the HDPE jugs and paperboard cartons that dominate the U.S. market. Attempts by Canadian LLDPE film producers to penetrate the much larger U.S. market with their form-fill-seal system have so far been unsuccessful. The U.S. public is apparently not ready for this energy efficient, low cost milk packaging system.

A relatively small volume of shelf-stable UHT milk is sold in the U.S. in a semi-flexible package that contains plastic components: the Tetra Pak Aseptic Carton. The major material of construction for this package is paperboard, but it also contains several layers of LDPE applied by extrusion coating to serve as seal layers.

The large institutional market for milk is served in part by the bag-in-box package described more fully below under "Beverages". The bag inside the box need only cleanly contain the contents, since the milk is refrigerated and even so has a short shelf life. LDPE/EVA coextrusions are strong enough and have the right seal characteristics.

Cheese

Unlike milk from which they are derived, both natural and processed cheese have a much longer shelf life, particularly when refrigerated. Most cheese products are packaged in composite multilayer films that incorporate barrier layers to restrict water loss, protect against airborne organisms, retard spoilage due to oxidation, and in some cases, prevent the access of ultraviolet light. Aluminum foil or metallized PET or OPP are used if barrier is more important than product visibility, but far more often, PVDC or EVOH perform this function so that consumers can also see the product.

Prior to retail packaging, natural cheese is aged for several months in EVA/PVDC/EVA shrink barrier bags like those used for subprimal beef cuts or in nylon/EVOH/EVA non-shrink constructions that do not distort the shape of the large blocks of aging cheese. The retail packages for natural cheese consist of: (1) either PVDC-coated 6 nylon or PVDC-coated OPP, both coated with EVA to provide a hermetic seal, (2) a PP/PE/PP/PVDC/EVA film, or (3) a thermoformed tray based on nylon lidded with PVDC-coated PET. In most cases, these packages are gas-flushed for additional oxygen protection. All these products are packaged on horizontal FFS machines, but the VFFS technique is used to package shredded natural cheese in PET or OPP bags containing PVDC barrier layers and EVA or ionomer seal layers.

These film constructions will not work for natural swiss cheese. It gives off CO_2 as it ages, and this gas buildup must be vented. This offers biaxially oriented nylon (BON) a rare food packaging opportunity. Fortuitously, BON has a very high CO_2/O_2 permeability ratio. This allows the CO_2 to be vented by diffusion without allowing the excessive inward passage of oxygen.

Processed cheese is sold in blocks or as packages of slices for sandwich making convenience. In one sliced cheese packaging system, molten cheese is cast directly onto the slice packaging film. PET coated with a PVDC-wax formulation is widely used in this process in spite of its high cost because of its excellent high temperature properties. Cold processed slices cast on steel drums are wrapped in PP/PE coextrusions. The stacks of slices from both processes are finally overwrapped in PET, OPP, or OPP/cellophane, all provided with one or more PVDC layers.

Special film constructions have been developed for the very large institutional cheese market: cellophane/OPP/PVDC laminations or PVDC/OPP/acrylic laminations predominate.

Other Dairy Products

Most of these—yogurt, margarine, ice cream, sour cream, and butter— appear in rigid packages. Sticks of butter are still wrapped in paper coated with wax or LDPE. Metallized PET is occasionally used for increased protection against light and moisture loss.

Dry Foods and Grain Products

Cereal

Dry ready-to-eat cereal must be protected against any change in moisture content and against loss of flavor components or gain of unwanted odors or flavors from the environment. Ready reclosability is a highly desirable feature, since the package contents are rarely consumed all at once.

The package that meets these requirements consists of a plastic film inside a paperboard box. The film is a two mil thick coextrusion of HDPE and EVA, with HDPE accounting for about 80% of the thickness. HDPE provides, at low cost, sufficient moisture barrier (about 0.3 g-mil/100 in²-day) along with adequate toughness to withstand puncturing by sharp-edged particles. In the rare cases where added ingredients such as fruit require additional oxygen protection, EVOH or nylon are included in the coextrusion.

Dry cereal packaging consumes about $50 million worth of HDPE and is a major food market for this film.

Baking Mixes

Like dry cereal, baking mixes need good protection against moisture and odor pickup, but the requirements are not as stringent as for cereal. An LDPE bag inside a box overwrapped with EVA-coated paper was the first package to use plastic film, but that has been largely replaced by a nylon/LDPE coextruded bag. This newer construction is lower in cost because it allows the elimination of the overwrap in which EVA served as an odor barrier, since nylon performs this function well.

Rice

For plain rice, the major package function is containment. Both paperboard boxes and plastic bags are used for this purpose, each of these packaging types having about 50% of the U.S. market. Plastic bags must be strong enough to contain up to 20 pounds of product, be puncture resistant and readily heat sealable. Both LDPE and LLDPE meet these requirements well at low cost and combinations of the two, with LLDPE used for additional strength, are the most commonly used packaging film. The construction is a two-layer laminate with the outer layer reverse printed and the inner layer performing the heat seal function as well as protecting the printing ink from oil or abrasion. Split peas and dried beans are also packaged in this way.

Flavored aromatic rice products demand a better odor barrier, which may be provided either by a PET/EVA pouch or by a vacuum brick package that uses a 4.5 mil PET/nylon/LDPE construction. One specialty flavored offering is packaged in cook-in bags perforated to allow access to the cooking water but not the loss of uncooked rice grains. These bags are made from PE blends, PP, or nylon, all of which prevent tear propagation from the perforations.

Pasta

About 2/3 of the U.S. pasta production is packaged in pouches made from two layers of adhesively laminated OPP film. OPP is chosen because its attractive optical properties are superior to less expensive films that also have the necessary mechanical properties and printability.

Dry Soup Mixes

These products must be carefully protected against moisture uptake and occasionally against oxygen. Most are packaged in flexible pouches that contain foil as the barrier layer. Popular constructions are:

- paper/LDPE/foil/sealant
- PET/LDPE/foil/sealant

The sealant is LDPE unless ionomer is needed for high fat content products.

Baked Goods

This $14 billion food category is a major market for plastic films. Bread and rolls have a 65% market share in this group of products which also includes sweet goods, cakes, pies, pastries and doughnuts.

Unless frozen, the shelf life of these products is only a few days, thus packaging is used only for moisture content control, cleanliness, and display.

Packaging of bread and rolls is dominated by the 1-1/4 mil LDPE bag made off-line by converters on pouch machines described in Chapter 5 and sold to bakers for filling and mechanical closure. This bag performs the necessary functions described above and is easily opened and reclosed. Specialty breads with lower moisture content are double wrapped for better moisture control. The inner wrap is variously waxed paper, cellophane, or OPP, but the outer wrap is always LDPE with a mechanical closure. No reclosure feature is needed for institutional bread, so it is wrapped and heat sealed in LDPE.

Cake doughnuts are packaged in OPP rather than LDPE because the packaging machines used have not been changed from the days when they ran cellophane, and thus do not handle well the limper LDPE. OPP is also used for sweet dough products and uncrusted fruit pies, both of which are more expensive than bread and rolls, strictly because it is more attractive—LDPE could perform this job equally well from a protection point of view. Small cakes use acrylic-coated OPP or a coextrusion with the structure: E-P copolymer/PP/EVA.

Frozen baked goods, which have only about 10% of the total market, are packaged in LDPE bags (bread), or in boxes overwrapped with OPP or polyethylene shrink film (cakes).

Cakes and other relatively expensive baked products are frequently packaged in boxes. Plastic films are used in most cases to serve as a window through which the contents can be viewed before purchase. Very high clarity, glossy PS films are typically used for these windows. Occasionally the boxes are overwrapped with OPP to extend shelf life by a few more days.

Snack Foods

As used here, this term embraces salty snacks, cookies and crackers, candy, gum and microwaveable popcorn. Next to meat and poultry, this is the largest food market for packaging films. It is dominated by multilayer constructions. Often these constructions include non-plastic components such as paper, foil, cellophane, or glassine, but plastic films have the largest share of this market.

Of that share, OPP has the largest piece. Snacks are the largest packaging market for OPP, consuming 25% of the total U.S. production. Its combination of good moisture barrier, relatively high modulus (fewer wrinkles), clarity, print receptivity, toughness, strength and low density cannot be matched by any plastic film that can be bought at an area cost even close to that of OPP.

Plain OPP has all these essential properties, but lacks heat sealability. Not only is its heat seal range quite narrow, but it also tends to pucker in the heat seal area due to localized relaxation of the film when it is heated. Several approaches are now used to confer good heat sealability on OPP:

(1) Modification with terpenes. While providing adequate sealability, the heat sealing range of these products is still narrow (25°F.) and the hot tack is poor. Nevertheless, it is often used for carton overwraps.

(2) Coating with an acrylic polymer. This widens the heat seal range to 80°F., imparts additional sparkle, and cuts the COF in half. This version is widely used on VFFS snack packaging machines and for cigarette overwrap.

(3) Coating with PVDC. This confers a wider heat sealing range (50°–80°F.) and increases the oxygen barrier by a factor of about 100. Film coated on one side with PVDC and the other with acrylic is lower cost, has enough barrier for many applications, and is more machineable than the two side PVDC-coated film.

(4) Coextrusion to form two or three layer films with a core of PP and outer layers of either ethylene-propylene copolymers, EVA, or

ionomer. Choice of the outer layer depends on the heat seal range desired, the degree of hot tack required, and cost. Ionomer layers are most expensive but give excellent hot tack and a heat seal range of about 100°F.

OPP's closest competitor is HDPE, which can be used where package contents are protected against light with pigments or inks or where its inferior clarity is not a marketing disadvantage.

Seal layers in these structures are usually LDPE, EVA, or ionomers, but sometimes PVDC, when needed for barrier, also doubles as a sealant. Today the seal layer is usually applied directly to the film base by extrusion coating or as a coextrusion component, so this is a minor market for these polymers in film form.

Although it is difficult to generalize about package requirements for a category as large as this one, there are some requirements that most of these products have in common:

- protection against moisture gain (and sometimes loss)
- good heat sealability on both vertical and horizontal form-fill-seal machines
- strength, and for some products, superior puncture resistance
- stiffness, to enable bags to stand up on supermarket shelves
- excellent print receptivity
- good runnability characteristics on high speed packaging machines
- attractive optical properties: clarity, gloss, sparkle
- good resistance to wrinkling after repeated handling
- low cost

Oxygen barrier, by and large, is not a crucial property for this product group except for those products, such as potato chips or cookies or some candies, that are high in fat content. For these, the combination of oxygen and UV light leads to light-catalyzed fat oxidation and rancidity. Thus foil is often used to block out both light and oxygen.

Salty Snacks

This application is dominated by multilayer films. The major component is either OPP or HDPE, with OPP currently dominant. PVDC, LDPE, and polyethylene copolymers (EVA, ionomers) are essential minor components used mainly to enhance barrier and heat seal characteristics. Both laminations and coextrusions are widely used. Laminations must be used when non-extrudable components such as paper, foil or cellophane are involved in the structure, and even more importantly, when the printing must be buried to protect its eye appeal. This latter reason will probably become the major driving force for laminations as continued resin improvement develops products that will perform the functions for which non-extrudable components are now used.

The most frequently used structures are:

- OPP/adhesive/HDPE/EVA or ionomer
- OPP/adhesive/glassine/PVDC
- Two coextruded OPP layers plus a seal layer. One OPP layer is an ethylene copolymer. The seal layer can be acrylics, PVDC, or ionomer, the latter providing higher machine speeds at higher cost.
- OPP/PE/metallized OPP/seal

In all these structures, OPP can be opacified by printing or pigmentation if a light barrier is needed.

HDPE coextrusions, typically HDPE/EVA or ionomer, are 20% lower in cost than the least expensive OPP structure, but so far, the inferior clarity of HDPE has limited its use to structures that are pigmented for light barrier or product concealment. As noted earlier, these or any other coextrusions also suffer the appearance drawbacks associated with surface printing.

The wide variety of structures used for salty snacks arises from the different package requirements of the various products in this category; potato chips need a light barrier, corn chips do not; fragile products need a stronger bag than do tougher products.

Cookies and Crackers

Like salty snacks, moisture barrier is a key requirement in this application. A moisture vapor transmission rate (MVTR) of 0.1 to 0.4 g-mil/100 in^2-day is usually needed. For products with high fat content, oxygen barrier is also needed to avoid rancidity but films with the necessary moisture barrier usually have an adequate oxygen barrier as well. Multilayer films are the usual choice to meet these requirements. Films of HDPE/EVA or OPP/E-P copolymer are used as the inner wrap for crackers packaged in boxes. A variety of coextrusions and laminations are used for bagged products or for box overwraps, but in all of these, OPP is the major component with PVDC, nylon, and EVOH providing the barrier and LDPE, EVA, or ionomer the seal layer. Paper and aluminum foil are frequently used components in plastic-based laminations, the latter in an attempt to achieve the ultimate in moisture, oxygen, and freshness control for soft cookies with a fresh baked flavor.

Candy and Gum

Chocolate candy requires an MVTR less than 0.5 g-mil/100 in²-day, an oxygen trasmission rate (OPV) less than 5.0 cm³-mil/100 in²-day-atm, a light barrier, and a seal layer that can be activated at temperatures close to room temperature. The OPP based structure:

Lacquer/ink/white OPP/PVDC/cold seal

dominates this application. The cold seal material is a water-based emulsion of a natural rubber latex. Metallized OPP is sometimes used as an alternate to PVDC for the needed barrier properties, particularly for candy containing nuts whose high fat content requires a lower OPV.

Bagged candy is packaged in OPP/LDPE bags or other constructions based on OPP. Gum, with each stick wrapped in foil laminated to coated paper, is bundled with an outer wrap that is a lamination of paper, OPP or cellophane, and adhesives.

Microwaveable Popcorn

This is the only snack category that requires packaging materials with high temperature capability that OPP does not possess. Packages for this product have evolved rapidly since its introduction in the 1970s. It was originally an important market for PET, but cost factors have recently led packagers to use paper as the main component, metallized PET as the susceptor (see "Dinners and Entrees" below) which converts microwave energy to heat to ensure complete popping, and a non-ovenable OPP overwrap to provide a flavor/odor barrier. Polyester survives as a major component only in products that are packaged in CPET bowls lidded with coated PET film.

Coffee

Whole bean coffee is universally packaged in paper bags with a liner of LDPE, PVC, or PET which serves only to protect the printed paper from staining by the greasy beans.

Today, most coffee is sold in the form of ground beans or soluble ground product. Most of the soluble varieties are packaged in rigid containers of metal, glass, or plastic, while about 50% of ground bean coffee is flexibly packaged. Ground beans require an excellent barrier component in their package. Foil or metallized films, either PET or nylon, are used for this purpose, combined with LDPE or ionomers for heat sealing.

Flexible packaging is used for soluble coffee in cafeterias and other food service establishments where single portions are provided in small paper/LDPE/foil/LDPE pouches.

Beverages and Other Liquid Food Products

The vast majority of these products are packaged in rigid containers made of metal, glass, or plastic, but even in this product category, plastic films have demonstrated their versatility by making significant inroads in two general areas: the plastic pouch for milk and the plastic pouch contained inside a paperboard box, the so-called "bag-in-box" package.

Bag-in-box packaging was introduced in the U.S. shortly after World War II as a convenient lightweight package for battery acid. The package, whose manufacture is described in Chapter 5, consisted then of a polyethylene bag fitted with a plastic dispensing spout and contained inside a paperboard box that protected the pouch against puncturing, provided a convenient way to carry it, and served as the billboard for the necessary graphics.

Since that time, the concept has changed little but has grown rapidly. This growth has been made possible by the introduction of new plastic films and converting techniques and has been driven by convenience, low cost, and the opportunity to present outstanding graphic displays on the box for the retail market. The convenience features are obvious: the package weight is a small fraction of the weight of the glass or metal container it replaces, the dispensing valve provides a simpler and safer way of removing a portion of the contents, and the easily flattened empty package is readily disposable. As noted in Chapter 5, the cost savings are impressive: a five gallon bag-in-box costs about 33 cents per gallon of contents while a 4.5 gallon steel can costs about 82 cents per gallon of contents.

The first food product to be packaged in this way was milk for institutional customers. LDPE again could serve as the bag material, since no barrier is required for short shelf life refrigerated milk and since the pouches are made off-line so the seal areas are not contaminated with the contents. As noted above, this is in sharp contrast to pouched milk packaged on a VFFS machine that demands a film such as butene copolymer LLDPE that will seal well through a film of milk.

The later development of EVA films, PET films, metallized films, and laminated combinations of these led to the expansion of this concept to many more food products and much larger packages, up to 330 gallons in capacity, for the institutional market to which most bag-in-box packaged products are sold. EVA has largely replaced LDPE as the dominant film for this application by virtue of its superior strength, flex crack resistance and heat seal capability. PET is used when high oxygen barrier is needed because it is readily metallized. Since the bags are never visible, they need not be transparent. In such situations, metallizing is an excellent low cost way of providing an oxygen barrier in the range of 0.06 cm³-mil/day-100 in²-atm. More moderate oxygen barrier in the 1.2–5 cm³-mil/day-100 in²-atm range is achieved in these constructions by using nylon rather than metallized PET.

Table 6.5. Applications of bag-in-box packaging.

Food Product	Quantity	Bag Construction & Other Comments
Dairy Products[a]		
Ice Cream Mix	2–3 gallons	PE/EVA bag, refrigerated storage,
Milkshake Mix	2–3 gallons	7–10 day shelf life
Frozen Yogurt	2–3 gallons	
Fluid Milk	5–6 gallons	PE/EVA bag, refrigerated storage, 7–10 day shelf life
Cream	55–300 gallons	PE/EVA. Refrigerated or ambient
Buttermilk	55–300 gallons	storage/transport
Anhyd. Milk Fat	55–300 gallons	
Wine		
Retail	2–6 liters	Met. PET/Polyethylene. Shelf life
Institutional	10–60 liters	approximately 9 months
Non-Alcoholic Beverages		
Apple, Grape, Citrus Juices and Fruit Punch	3–6 liters retail, 10–20 liters institutional	Met. PET/Polyethylene. Can be aseptic or non-aseptic
Soft Drink Syrup	10–20 liters	PE/EVA/Nylon
Juice Concentrates		Met. PET/Polyethylene. Diluted at point of sale
Tropical Fruits[b]		
Pineapple, Papaya, Guava, Mango, Banana, Passion fruit, Coconut Cream	3–6 gallons institutional, 55–300 gallons industrial bulk	Aseptically processed, Met. PET/Polyethylene, single or double layer
Temperate Fruits[c]		
Peach, Pear, Apple, Apricot, Berries	1–6 gallons institutional, 55–300 gallons industrial bulk	Aseptically processed, Met. PET/Polyethylene, single or double layer
Tomato Products[d]		
Catsup, Pizza Sauce, etc.	2–6 gallons institutional, 55–300 gallons industrial bulk	Aseptically processed, Met. PET/Polyethylene. (High acid content allows use of moderate barrier construction vs. low acid fruits, above.)

(continued)

Table 6.5. (continued).

Food Product	Quantity	Bag Construction & Other Comments
Formulated Fruit Products		
Fruit-based toppings	2–6 gallons	Met. PET/Polyethylene, single or
Ice Cream	institutional,	double layer depending on contents.
Yogurt	55–300 gallons	Usually aseptically processed.
	industrial bulk	
Condiments		
Mayonnaise	2–6 gallons	Met. PET/Polyethylene. Aseptic or
Salad Dressings	institutional,	hot fill, depending on contents and
Edible Oils	55–300 gallons	shelf life requirements
Mustard	industrial bulk	
Egg Products		
Liquid Whole Egg	3–5 gallons	Nylon/PE or Met. PET. Aseptically
Salted Egg Yolk	institutional,	processed and refrigerated for
	55–300 gallons	storage and transport
	industrial bulk	
Other Low Acid Foods		
Soups, Sauces, Stews,	3–5 gallons	Met. PET/Polyethylene, single or
Coconut Milk, Gravy,	institutional,	double layer. Aseptically processed
Carrot Paste	55–300 gallons	
	industrial bulk	

[a]The shelf life of yogurt and the mixes can be extended to 45 days by aseptic processing and packing followed by refrigerated storage and transport.
[b]As purees or crushed pineapple or pineapple juice concentrates.
[c]As purees, pulp, diced, or sliced.
[d]As purees, paste, crushed, or diced.

Table 6.5 illustrates the variety of products, sizes, and constructions found today in this package concept. It also shows why a good oxgyen barrier is now an important requirement for many bag-in-box packaging applications.

Since the Tetra Brik or Brick-Pak described above in the section devoted to milk is not a true flexible package and since it involves plastic resins but not plastic films in its construction, its widespread use in liquid food packaging, particularly outside the U.S., will not be described here in any detail. It will suffice to point out that when it is used for liquid foods other than milk, an aluminum foil oxygen barrier is normally incorporated in the structure. This requires the use, in addition to LDPE, of ionomer or acid copolymer layers to adequately bond the foil in an aggressive acid juice environment to the rest of the structure.

Cooking Oil

In the U.S., all cooking oil is packaged in rigid containers made of glass or plastic. However, in many populous but relatively poor areas of the world where cooking oil consumption is much higher per capita than in the U.S., cooking oil is dispensed at retail from metal drums into refillable containers brought to the store by the purchaser. This practice enables the oil seller to increase profits by diluting or adulterating the product with less expensive and less sanitary materials. Deaths have been reported in Spain, India, the Phillipines, and several Latin American countries as a result of this regrettable practice. Since the cost of prepackaging oil in metal or glass would be prohibitively high in these low-income regions of the world, governments, particularly in India, have begun to support the development of inexpensive plastic pouches for this purpose.

This is a packaging application that resembles pouched milk except that here an oxygen barrier and sealability through a film of oil are both required. In addition, the pouches must be very strong to withstand the rigors of transportation over sometimes primitive roadways.

LLDPE, which works well for milk, will not meet all the barrier requirements. Low cost coextrusions of HDPE/acid copolymer or HDPE/adhesive/nylon/adhesive/acid copolymer have good seal integrity and sufficient barrier for a 90–120 day shelf life.

This application is still highly developmental, but the huge potential market for plastic films it offers guarantees that cooking oil in plastic pouches will eventually be widespread in many countries outside the U.S.

Spreads

The single serving package for mustard, ketchup and salad dressing found in all food service establishments, airlines and hospitals is a complex multilayer structure that must provide oxygen and moisture barrier, reliable hermetic seal capability and good puncture resistance. Two structures are used that meet these requirements:

LLDPE/tie layer/foil/tie layer/PP
and
OPP/white LDPE/tie layer/olefin copolymer/acid copolymer

The latter structure is increasingly used for high acid products such as ketchup and mustard that do not need high oxygen barrier. It avoids the leakage problems sometimes encountered with the former.

Ice

Although not normally considered a food, packaged ice is nevertheless an important consumer of plastic film. As Table 6.1 shows, about 30 million pounds of EVA film are used for this application. Producers would no doubt prefer to use low cost LDPE, but the greater strength and superior puncture and tear resistance of EVA requires the use of this slightly higher cost material when packaging heavy 25 pound blocks or bags of ice cubes.

Dinners and Entrees

The enormous growth of precooked dinners and entrees designed to be quickly heated and then consumed is a well-known phenomenon that needs no elaboration here. The container for these products generally consists of a plate or bowl made variously of plastic-coated paper, thin aluminum sheet, or more commonly of a thermoformable plastic such as PET or nylon that can withstand oven heating to 400°F. without distortion and without transfer of its components to the food. Plastic films are the most common material used for the lids on these trays or bowls, since the aluminum foil lids used on the original versions of these packages are not compatible with the ubiquitous microwave oven. In fact, were it not for the huge popularity of the microwave oven, it is likely that aluminum foil, with its excellent barrier properties, would still dominate the lidding application for these products.

Since the package usually remains lidded while being heated, the high temperature capability required for the plate or bowl generally extends to the lid as well. A dual structure consisting of aluminum foil capped with a thermoformed HDPE lid was a popular early design, but PET film, coated with a special copolyester seal layer, is becoming a more common choice. This PET film lid has the advantage that it can be draped over the product and heat-shrunk to follow the product contours, thus minimizing dessication in the freezer ("freezer burn").

For dinners and entrees that are sold refrigerated or at room temperature rather than frozen, oxygen barrier becomes important. For refrigerated products, PET of sufficient thickness will serve, particularly when a shelf life in the range of 30 days is all that is needed, but very long shelf life unrefrigerated products that are still very new on the market must use aluminum foil or have a PVDC or EVOH layer incorporated in the lid as well as in the tray or bowl.

Microwave-heated dinners lack certain esthetic taste and appearance features found in meals cooked in radiant ovens: products that should be browned and/or crisped are not, and it is hard to cook various meal components selectively. These drawbacks have led to another film application for this product category: the so-called microwave susceptor. This commonly consists of a piece of metallized PET film laminated to a piece of paperboard and inserted in the package. It converts microwave to radiant energy, thus more or less duplicating, from the food point of view, the temperature environment found in the radiant oven. Optimum engineering of these susceptors is a complex and as yet poorly understood art which is beyond the scope of this book.

A totally different flexible packaging development for these products took place in parallel, chronologically speaking, but has never become as popular as the tray-lid package. This is the high-barrier film bag, commonly called the "retort pouch" because the food is packaged in the pouch and then retorted, as is canned food, at about 250°F. This package was developed by converters seeking additional markets for aluminum foil. They visualized a shelf-stable package with two major advantages over the metal cans or glass jars: light weight, for low cost and easy portability, and a thin cross section.

The latter feature allows shorter retorting times to be used. This avoids overcooking the outer food layers since the thickness through which heat must be transferred to adequately cook the center is far less than in a cylindrical container. When microwaveability became important, non-metallic barrier layers based on PVDC were incorporated in place of the aluminum foil.

The successful development of this package was a major technical achievement. Seal polymers that could seal through layers of food and then withstand retorting conditions and still remain hermetic for a year or more and constructions that would have excellent oxygen barrier but would not contribute undesirable components to the food under retort conditions were the major hurdles that had to be surmounted.

The original foil-containing structure was:

PET/adhesive/foil/primer/adhesive/PP sealant

The microwaveable embodiments are:

PET/PVDC/PP,

or the more complex

nylon/PVDC/nylon/tie layer/sealant.

For a variety of reasons, this package has never caught on with the buying public in a major way. It survives mainly by virtue of its use by the U.S. Army in its MRE ("Meals-Ready-to-Eat") program, which consumes over 25 million meals in retort pouches each year.

Pet Food

Dry pet food requires a grease-resistant package that will hold the product in quantities up to 25 pounds, keep it clean, and provide a surface for multicolored advertising messages. Bags are the most popular way of meeting these requirements. The construction most widely used consists of lacquered paper lined with PP or PE, either laminated or as a separate layer, for heat sealing, abrasion resistance and a grease barrier.

Some semi-moist pet food products, which have a moisture content less than 35%, are flexibly packaged in bags that must provide moisture, oxygen and aroma barriers. This demands a multilayer structure, and several are used:

paper/PE/foil/ionomer

paper/tie layer/metallized OPP/ionomer

Ionomer resins are needed here to allow sealing through fat and grease. Tie layers are either LDPE or acid copolymers. Foil or metallized OPP provide the needed barrier characteristics.

CONCLUSIONS

This discussion of the many uses of plastic films for packaging food is appropriately concluded with a few general observations.

The availability of plastic films for packaging foods has made the act of packaging itself far more prevalent, since only plastic films can provide at low cost the many types of protection demanded by perishable foods. As low cost food packaging has become more and more widespread, food spoilage has declined, allowing the arable areas of the globe to continue to support a geometrically increasing population.

The savings in weight made possible by packaging in plastic films lead to enormous savings in energy. These energy efficient packages have significantly extended the world's supply of non-renewable fossil fuels. Of all flexible packaging materials used, plastic films are the most energy efficient, in the sense that the ratio of fuel saved to fuel consumed in producing the weight of packaging material required is the highest for plastic. This subject is discussed in more detail in Chapter 7.

The advent of transparent films, beginning with cellophane, revolutionized the food packaging and marketing process. Supermarkets appeared in which products could be displayed and selected without the intervention of a salesperson. Impulse buying of attractive looking items, sometimes rack-hung near the checkout counter, became a powerful factor leading to elaborate graphics that either supplemented or complemented the appearance of the products themselves. Plastic films, following cellophane, greatly expanded the scope of this revolution by providing many important features and functions that cellophane did not possess—usually at lower cost.

There are innumerable examples of packaging, distribution and merchandising techniques that became available to the food packager with the advent of plastic films. Here are a few:

- Processed meats can be centrally packaged, lowering cost, improving quality and providing ready brand identification.

- Precooked dinners and entrees can be offered ready to heat and eat with little effort and minimum cleanup. This revolution in dining has accompanied, and in some senses facilitated, the major life style changes that have taken place in the years following World War II.
- Bag-in-box packaging saves weight and allows the packaging of items that can be dispensed a little at a time without the concomitant introduction of air which will spoil the undispensed remainder of the contents.
- Fresh produce can be packaged in controlled atmospheres that maintain the CO_2/O_2 ratio at the optimum level for the item being packaged and greatly extend the shelf life of these products.
- The retort pouch saves weight and improves the taste of retorted foods by avoiding overcooking during retorting. While not very large in its own right, the retort pouch can be regarded as the exemplar of the inevitable trend from metal and glass to plastic packaging of oxygen-sensitive foods.

This list could be extended almost indefinitely, but doing so would be excessively redundant in view of the information already provided in this chapter.

PACKAGED RETAIL PRODUCTS OTHER THAN FOOD

In this highly diverse product category, whose value totals approximately $400 billion annually in the U.S., the role of plastic films in packaging is quite different than it is for the food products discussed above. Only an insignificant fraction of non-food products require protection against moisture, oxygen or odor transfer. Tobacco products, perfumes, pharmaceuticals and cosmetics are examples of those that do. For the vast majority, the principal package functions are:

- to contain the product
- to protect it against dust and dirt
- to display the product in a way that enhances sales appeal and advertises the contents and the manufacturer
- to assemble many small items, bundle different but related items together or bundle several identical items together for a multiple sale
- to discourage pilfering or tampering

While this is by no means a short list of essential functions, the conspicuous lack of gas or light barrier requirements means that the necessary functions can usually be satisfactorily performed by a single layer of film.

Thus, as compared to food packaging, the packaging materials used in this category are simpler and less expensive, the multilayer film expertise of the flexible packaging converter is rarely required and often the packager buys film directly from resin manufacturers who also make film, thus bypassing the converter.

Tobacco Products

Cigarettes are universally double-wrapped. The inner wrapper that surrounds the cigarettes is a non-plastic structure consisting of aluminum foil adhesively laminated to bleached kraft paper. This inner wrap is not sealed but nevertheless provides a degree of moisture protection. The package is then overwrapped in plastic film.

Cigarette and cigar package overwrapping became the exclusive domain of cellophane once that transparent film, which both protected and enhanced the eye appeal of these products, became available. When functional seal materials such as acrylic polymers were successfully developed for OPP, that lower cost film took over this application, which requires a good moisture barrier, high clarity and gloss and good heat sealability and machinability on the very high speed overwrapping machines that have been developed specifically for cigarettes. For a time, PET with a heat sealable PVDC coating held a minor share of this market because its greater stiffness resisted wrinkling better than OPP. Packagers finally stopped paying the 20–30% premium that producers demanded for this more expensive film whose value in industrial applications fully justified its higher price. As shown in Table 6.1, cigarette and cigar package overwrap is now the third largest market for OPP in the U.S.

Pipe tobacco is packaged in metal cans, glass jars and flexible pouches. The latter have one consumer appeal feature not possessed by the others: the retained moisture content, and thus freshness, of the contents can be assessed by squeezing the pouch. Since pipe tobacco packages must be opened and closed many times before the contents are exhausted, and since the high moisture content and volatile aromatic ingredients are very important aspects of the product taste, the requirements for moisture and odor barrier and near-hermetic reclosability are much higher for these pouches than for cigarette packages. Thus aluminum foil is usually the barrier component in the pouch. The entire structure is glassine/LDPE/foil/LDPE. The glassine carries the printing while LDPE bonds foil to glassine and provides the inner seal layer.

Inexpensive pipes are frequently skin packaged in PVC or ionomer for carded display; lighters and pipe filters are blister packaged in PVC.

Perfumes, Toiletries and Cosmetics

These products are generally formulated with high moisture content and contain vital, oxygen sensitive odor ingredients. Although plastic containers could no doubt be successfully used to package them, product image demands that glass or decorative metal containers be used. Plastics nevertheless play an important packaging role, since many of these products are sold on a self-service basis in drug stores and supermarkets where they are blister packaged with PVC on a printed card and displayed by hanging on racks. Closely related personal grooming products such as hair curlers, barrettes and so forth are blister packaged for sale in the same way. A few of these products, such as disposable razors, are packaged in groups in rack-hung LDPE bags.

Hardware and Housewares

This product category is one of the major domains of the blister package, the skin package, and the small rack-hung bag. Supermarkets, hardware stores, drug stores and other retail merchants of these products have almost universally adopted hanging racks as the most convenient and space-efficient way to stock and display small items. On these racks, carded PVC blister or ionomer skin packages most effectively provide the necessary functions of product containment, display, protection and pilfering discouragement in a glossy package that also serves to communicate product information and advertising. Lower cost LDPE rack-hung pouches, also widely used, can contain the product and keep it clean, but are inferior in eye appeal and do not communicate as effectively with the shopper—more often, the pouched contents must essentially sell themselves.

Boxed Products

Boxed items are often sold without a plastic film overwrap. When they are overwrapped, the choice is OPP or shrink wrapping in high clarity shrink film, usually a polypropylene copolymer or PVC. High cost, high clarity, stiff films dominate this subcategory because eye appeal on the shelf is regarded as an important sales tool and a neat, wrinkle-free package is essential.

Large Unboxed Products

Large unboxed items, particularly those with awkward shapes, are sometimes bagged in LDPE, sometimes shrink wrapped in PVC or polyolefin shrink films, but are often sold unwrapped. There are two motives for overwrapping these kinds of products. Items such as paint roller sleeves, rolls of paper towels, paper napkins, wallcovering, drop cloths, lawn chairs and mop heads, all of which should be kept clean before use, are often overwrapped, usually in LDPE bags. Items such as large wrench sets and many other tools are blister or skin packaged to prevent loss by pilfering.

One large, awkwardly shaped housewares item is worth mentioning, not only because it is a tribute to the versatility of shrink packaging but also because of its humorous aspect: the shrink packaged toilet seat. It is difficult to imagine a valid motive for this packaging exercise—cleanliness, perhaps, or possibly added eye appeal, although the latter motive seems incongruous in view of the utilitarian nature of the packaged item.

Textile Products

Soft textile products (sheets, towels, blankets, quilts, pillows, drapes, sleeping bags, etc.) are often sold unpackaged. This is particularly true for towels and bathmats where the customer wants to feel the product. More expensive blankets and quilts are packaged in thick calendared PVC film provided with gusseted ends and a zipper closure for product storage when not in use. Sheets, which are a relatively dense product with a uniform shape, are frequently shrink wrapped in high gloss PVC or polyolefin films. The distorting effect of shrink wrapping does not affect a dense product of this kind. Alternatively, unoriented polypropylene is sometimes used for sheets and blankets to provide a high gloss package where the greater strength of OPP is not needed. Pillows and sleeping bags, when they are packaged, are universally contained in large heat sealed LDPE bags or pouches that loosely contain the product and serve to keep them clean. Manufacturers apparently regard the low cost of LDPE bags for these products as more important than the superior eye appeal that could be provided by heat sealable PP or PVC.

Apparel

Few apparel items are overwrapped for store display. Exceptions to this rule are panty hose, men's dress shirts, pyjamas, underwear, and occasionally socks, all packaged in LDPE bags. All other apparel items are displayed on hangers. The need to closely examine and try on apparel makes overwrapping in the store impractical for most of this large retail product category.

Soiled clothing cleaned professionally is always protected from dust and dirt with an LDPE bag before it is returned to the customer.

Kitchen Wrap

This is one of the very few applications for PVDC film used by itself rather than as a component of multilayer constructions. Although it is an expensive film, its unique combination of clarity, gloss, excellent barrier, limpness and high cling make it the plastic that competes most effectively with aluminum foil in this application.

Stationery and Paper Products

This application category includes such items as paper tablets, bundles of individual sheets, envelopes, bundles of index cards and note cards, wrapping paper and gift wrap. Often these products are sold unwrapped. When used, a plastic film overwrap serves as a dust cover and as a convenient way to bundle several items together. The principal criterion that determines whether these products are overwrapped is the stiffness of the contents. This is because shrink wrapping in PVC or polyolefin shrink film is the least expensive and thus the dominant wrapping style for these products but the shrink force of today's polyolefin shrink film is so high that it badly distorts a thin pile of paper. On the other hand, piles of cards or wrapping paper wound around a tubular cardboard mandrel are stiff enough to resist deformation in the shrink wrapping process. Low shrink force polyolefin films are being developed for this large shrink film market.

Toys and Games

These products are sold in carded PVC blister packages and in boxes that are either unwrapped, overwrapped in heat sealable OPP or shrink wrapped in PVC or polyolefin shrink films. Here the package serves to protect against dust and dirt, to discourage pilfering, to bundle several related items together, to add sparkle and eye appeal to the contents and to communicate with the customer. Thus the packager demands a clear, attractive film that can be applied at low cost on high speed packaging machines. OPP and shrink films both meet these requirements well but shrink films are often less expensive, making this a large market for them.

Like boxed toys, records and tapes are always overwrapped for cleanliness and tamper evidence. They are stiff, uniform in shape and contain colorful graphics that must be highly visible at the point of sale. Those attributes make this an ideal application for clear PVC and polyolefin shrink films, whose low per-package cost allows them to dominate this market. Audio accessories sold along with these products are carded in PVC blisters or skin packaged with ionomer or LDPE film.

The discussion above reinforces the observation made at the beginning of this section: the plastic film packaging of this multitude of products serves the purposes of containment, cleanliness, security against tampering and product advertising. Barrier characteristics are rarely important. These objectives can be met with simple monolayer films coated if necessary for heat sealability. LDPE, PVC, and coated OPP dominate these applications. LDPE usually makes the lowest cost package and is found mainly in pouches; PVC is the best thermoformable plastic for clear blister packaging; and OPP is the lowest cost stiff film for use when attractive shelf appearance is an important objective, where it often competes with the equally attractive PVC and polyolefin shrink films.

HEALTH CARE PACKAGING

The number of products involved in health care is enormous in size and diversity, ranging from operating tables to disposable surgical swabs. Most of these products are packaged in quite conventional ways and a thorough treatment would leave the reader with little new information. This discussion therefore focuses on two areas that differentiate packaging in this product category from the others in this book: small devices that must be sterilized after packaging, and pharmaceuticals where the unit dose concept provides a major new market opportunity for plastic films and sheet.

Sterile Device Packaging

The primary function of the package for these items is to allow sterilization to take place and then to maintain sterility of the device within. Thus the package must have good strength and puncture resistant properties, must prevent the intrusion of microorganisms from the environment and must be compatible with the sterilization method to be used. The latter requirement means that the package must be:

- able to withstand steam at 212°F. or dry heat at 325–350°F. for thermal sterilization
- sufficiently porous to allow passage of ethylene oxide if gas sterilization is used
- resistant to degradation by gamma rays, if radiation sterilization is used

Low cost radiation sterilization is now used for over 60% of packaged medical devices and will probably become the dominant technique in the future. To enable radiation sensitive plastic films to compete for this market, film producers have developed radiation resistant grades of common films such as PVC, PP and acrylic copolymers to allow those materials to compete with PS, PET and paper, which are inherently highly radiation resistant.

All these plastic films are now used for this packaging application, but the versatility of plastics does not manifest itself here because the high temperature resistance required to withstand thermal sterilization conditions and the porosity required to facilitate access of steam or chemical sterilants are not normal attributes of most plastic films. Thus many medical device packagers have tended to use more thermally resistant materials of controlled porosity, such as paper or spunbonded polyethylene, in conjunction with rigid containers, often thermoformed from plastic sheet, rather than the plastic bags or pouches that now dominate so many other packaging applications. Thus the sterile medical device packaging market is large (300–400 million pounds) for *plastics* but consumes less than 30 million pounds of *plastic films*.

The most common plastic packaging technique for medical devices is the rigid thermoformed tray made of PVC, PS or PETG. If these trays are lidded with plastic film, which is possible only if radiation sterilization is used, the films of choice are usually PVC or less commonly LDPE. Coatings are frequently used to provide a clean-opening, peelable seal.

Large items are flexibly packaged in bags made of paper, spunbonded polyethylene or plastic films such as LDPE, OPP, PET or combinations thereof equipped with paper or spunbonded polyethylene vents to allow the passage of gas or steam. If the items to be packaged are not compatible with the common sterilization methods and must be aseptically packaged, these same films are used, either alone or as multilayer constructions.

Drugs and Pharmaceuticals

About 70% of these products are solids – tablets, capsules, or powders – and the balance are liquids or gels.

Solid Pharmaceuticals

About 50% of the solid products are now packaged in jars or bottles and 50% in flexible materials. The fraction packaged flexibly is growing at the expense of the former, primarily due to the increasing popularity of unit dose packaging, the fastest growing type of packaging in the medical field. This packaging method gives better product protection, surer control over the quantity dispensed and positive identification of the product being dispensed. Over 90% of the unit dose packages are in the form of blister packages, which consist of a plastic film web with thermoformed blisters at intervals along its length and a flat web sealed to it. The flat web may be foil, paper, or multilayer plastic/paper/foil constructions such as PVC/PE/PVDC, depending on the degree of barrier required to protect the contents. The blister is flexible enough to allow the tablet to be pushed through the thin flat web by pressing on the blister.

PVC dominates this application on the clear side with close to 100% of the volume, which now amounts to roughly 18 million pounds per year. The dominance of PVC stems from the same factors that make it popular for blister packaging other products: clarity, strength, thermoformability, barrier properties and low cost.

Although unit dose blister packaging began with ethical products, it is now spreading to over the counter (OTC) pharmaceuticals where its higher cost is offset by the convenience and safety features noted above. Although only 20% of all oral solid dosage forms are now blister packaged, new solid drug packaging investment is weighted to the unit dose blister package concept which should be dominant in another 10 years.

Solid drugs in powdered or granular form make up only about 10% of the solid drug volume and are usually packaged in plastic film pouches using VFFS filling and pouch-forming processes. Again PVC has the major share of the market—about 65%—with the balance shared by OPP, HDPE, and some composite films. Flexible packaging of these products is becoming the dominant mode, as compared to rigid, for the usual reasons of light weight and low cost.

Liquid Drugs and Pharmaceuticals

Compared to solid products, drugs in liquid form make up about 1/3 of the total drug volume. Rigid containers such as tubes, ampules, jars and bottles are used for about 90% of these liquid products, but plastic pouches, sachets, and strip packs are growing in popularity. Such flexible, low cost containers are ideal for unit dose dispensing and can be constructed with several compartments, allowing the various ingredients to be mixed immediately before use. IV solutions, whole blood, and blood components are frequently packaged in flexible PVC pouches.

A variety of plastic films are used for this application: LDPE, HDPE, OPP, and PVC all share the monolayer film segment, while barrier films based on LDPE or OPP with EVOH or PVDC are used as needed. Strip packs for unit dose dispensing of liquids are usually made from PVC for the same reasons as noted above for solid drugs.

Tamper Evident Drug Packaging

Plastic films have become a common component of the total package used for solid and liquid OTC drugs even when the primary package is a rigid bottle or jar. Since these products are outside the control of the pharmacist, some tamper evident system is always incorporated in these packages. These are usually either a foil membrane across the vial lid, a PVC shrink band around the vial cap, or a plastic film overwrap around the box. Sometimes more than one of these systems is used in the same package to present additional deterrence to a tamperer.

TRASH BAGS, MERCHANDISE BAGS AND GROCERY SACKS

In contrast to the medical device market, this application category rivals food packaging in the volume of plastic film it consumes. It is the largest single film application for HDPE (54%) and LLDPE (45%) and consumes 14% of the total LDPE film. The percentages are high for HDPE and LLDPE and much lower for LDPE for two reasons: (1) the former two polymers outperform LDPE in this application, and (2) LDPE is a better low cost choice than the other two for many non-food items and large volume food products such as produce.

Most texts on packaging do not consider these bags as packages. This strikes us as short sighted. Not only do these bags perform the most ancient and essential function of a package—containment—but the sheer size of this market for certain plastic films demands that it be discussed to give the reader the best overall perspective on packaging film applications.

As noted, these packages perform only one of the many packaging functions described in Chapter 4. Because of this simplicity, the continuing evolution of this market among resin types has been driven largely by cost. Since these three resins are all priced close to one another, the cost issue is often a film strength issue: the stronger the film, the thinner it can be, and thinner film means less resin consumed per bag. Since the resin cost is well over 50% of the total cost of these bags, a lower per-bag resin cost is the most important criterion for selection among the three resins used. Table 6.6 abstracts the pertinent figures from Table 6.1 for ready reference in the discussion that follows.

These figures represent a market in transition. Most of this market will be shared between LLDPE and HDPE, with the two competing across the

Table 6.6. Consumption of resin types in bags (pounds in millions).

Resin	Trash Bags	Grocery/Merchandise Bags	Total
HDPE	139	306	445
LDPE	372	105	477
LLDPE	894	198	1092

board for all three bag types. LDPE persists in drawstring bags and in some merchandise bags where a soft film is desirable. LLDPE can be lower in per-bag cost than LDPE because its molecular structure makes LLDPE film about twice as strong as LDPE film. High molecular weight HDPE, a relatively recent development in the U.S., has even greater strength along with tear and puncture resistant properties sufficiently superior to those of LLDPE to make it the choice for grocery and merchandise bags. Trash bag makers do not require these properties to the same degree and can make their bags equally well from LLDPE at a lower cost per *pound*. Nevertheless, the greater strength of high molecular weight HDPE allows sufficient downgauging, vs LLDPE, to make it quite competitive on a cost per *bag* basis, even in the trash bag application.

This oversimplified picture ignores the realities of day to day price competition between these three resins. The actual resin price paid by large volume bag makers is always lower than the list price and fluctuates as resin manufacturing capacity and supply change over time. Nevertheless, the future should see increasing dominance of the grocery and merchandise bag segment by high molecular weight HDPE and stiff competition between LLDPE and high molecular weight HDPE for the trash bag segment.

HEAVY-DUTY BAG PACKAGING

About 10% of the heavy-duty bags are all-plastic in construction and are used rather than paper to contain large quantities (25–50 pounds) of products that are either moisture sensitive or must be protected against weather. Fertilizer, mulch and other lawn care products, hygroscopic plastic resins bagged for shipment to industrial customers, rock salt, and insulation batts and other building products exposed to weather at construction sites are common examples of products contained in plastic rather than less expensive paper bags because of the additional protection they require or because they cannot always be stored under cover.

These bags are produced from tubular or flat film. About 60% are single ply, made from LDPE, LLDPE or blends thereof. The much greater strength of LLDPE has allowed downgauging in these bags by almost a factor of two from the 4 mil thickness originally used. Multiple ply bags are either LDPE/LLDPE or LDPE/LLDPE/HDPE coextrusions, with HDPE used in the structure when additional heat resistance in hot filling operations is needed. For maximum strength, spunbonded polyolefins or a special cross-laminated HDPE known as "Valeron" are used. For example, pumpable gelled explosives are chub packaged on Kartridge-Pak machines using Valeron film.

Most heavy-duty plastic bags are preformed, but the form-fill-seal concept widely used for lighter weight packages is being increasingly used for heavy packages where the contents are free-flowing, such as hygroscopic plastic resins.

Although it is not, strictly speaking, a heavy-duty bag application, plastic film is often used as an inner liner for metal drums. The film serves to isolate the contents from the drum, protecting the drum from soiling and corrosion and protecting the contents from contamination by dirt or corrosion products on the drum. LDPE and LLDPE are the films most commonly used, since they provide the necessary strength and chemical inertness at lowest cost.

PALLET WRAP AND STRETCH WRAP

Bags or boxes stacked on pallets must be unitized, or held firmly in place, during transportation and warehouse handling. The classical approach was steel or plastic strapping, but these methods have now been largely replaced by shrink wrapping in heavy-duty LDPE shrink film and by stretch wrapping in EVA or LLDPE stretch film.

Pallet shrink wrapping was faster and gave better environmental protection than strapping and also was more secure because the entire load was surrounded by heavy plastic film. Because the load was evenly surrounded, the need for corner boards to prevent the load being cut by straps was eliminated.

The advent of high strength, stretchable EVA film led to the widespread use of stretch wrapping instead of shrink wrapping. Today LLDPE, with the same strength, cling, and stretch properties as EVA but 20% lower in cost, has become the dominant pallet stretch wrap film.

These stretch wrap concepts are discussed in the machinery context in Chapter 5, where it is pointed out that the availability of LLDPE and EVA films that can be stretched to 300% of their original length not only greatly reduces the film cost per pallet by at least a factor of five, as compared to shrink wrapping, but also consumes far less energy (no shrink tunnel) and makes a more tightly bundled and thus a more secure load. It is also a more versatile bundling system: a wide variety of load shapes and sizes can be wrapped at a variety of wrap tensions on a single machine from a single width of film, greatly simplifying equipment and film inventory requirements.

Important properties that a successful stretch film must have are very high elongation, minimal neck-in when stretched, high ultimate machine direction tensile strength, high puncture strength, low stress relaxation, high elasticity, high creep resistance and resistance to tear propagation, and good cling. The latter enables the operator to complete the wrap after cutting the film by simply wiping the loose end firmly in place. No adhesives or mechanical fasteners are required. EVA with 7% VA content has inherently good cling, but LLDPE must be formulated with tackifiers to match EVA in this important property.

The versatility of stretch wrapping has led to its use for wrapping and bundling many other items in addition to pallet loads: bulk shipments of bottles, building products such as doors, window frames, and foam insulation, automotive parts and garden implements, and large rolls of carpeting, paper and aluminum.

A decade ago, stretch wrapping was one of the largest markets for EVA film, but LLDPE now holds at least a 75% share of this 200 million pound market and probably will eventually displace EVA entirely.

BIBLIOGRAPHY

Bakker, M., ed. 1986. The *Wiley Encyclopedia of Packaging Technology*. New York, NY: John Wiley and Sons, Inc..

Benning, Calvin J. 1983. *Plastic Films for Packaging*. Lancaster, PA: Technomic Publishing Co.

Briston, J. H. 1989. *Plastics Films*. Harlow, England: Longman Scientific and Technical, 3rd edition.

Brody, 1970. *Flexible Packaging of Foods*. Boca Raton: CRC Press.

Griffin and Sacharow. 1975. *Drug and Cosmetic Packaging*. Park Ridge: Noyes Publications.

Hanlon, J. F.. 1984. *Handbook of Package Engineering*. New York: McGraw Hill.

Jenkins and Harrington. 1991. *Packaging Foods with Plastics*. Lancaster, PA: Technomic Publishing Co.

Sacharow, Stanley. 1976. *Handbook of Package Materials*. New York: Wiley.

Sacharow, Stanley and Aaron L. Brody. 1987. *Packaging: an Introduction*, Orlando: HBJ Publications.

Sacharow, Stanley and Roger C. Griffin. 1973. *Basic Guide to Plastics in Packaging*. Newton, MA: Cahners Books.

1990. *U.S. Pharmacopoeia and National Formulary*. v. 22.

CHAPTER 7

Future Trends and Possibilities

INTRODUCTION

The future for any product such as plastic packaging films is governed mainly by evolving trends in society, economics, technology and governmental action. This chapter is divided into sections that focus on each of these trend areas.

GOVERNMENT ACTION

The rapid growth in plastic packaging since World War II has led to a concomitant increase in the amount of plastic packaging waste that is discarded. Today, plastic waste of all types now accounts for about 7% by weight and about 18% by volume of the 160 million annual tons of waste that must be handled by municipal waste disposal systems. Of the total plastic waste, the largest fraction is plastic packaging waste. About 60% *by weight* of the plastic packaging waste consists of rigid plastic containers such as PET beverage bottles, HDPE milk jugs and EPS fast food take out trays, but the *volume* percentage of these rigid items is over 80%, far higher than the weight percentage. These figures apply to waste after compaction in landfills. The dominance in volume percentage of rigid plastic containers in uncompacted household trash is even greater.

The U.S. now faces a waste management problem which has reached crisis dimensions due to widespread closing of landfills in certain heavily populated states along the East Coast. As the public and their elected officials struggle to cope with this problem, their attention has focussed on the plastic component of the waste stream. Even though this component is minor, both by weight and by volume, it is not only highly visible, particularly in uncompacted household trash, but also it is regarded as uniquely offensive because of the widespread belief that it is the only component for which there is no solution other than reduction in consumption. Unfortunately for plastic *films*, the prominence of low density, high volume, rigid plastic *containers* in household waste has led the public to attack without discrimination *all forms* of plastic packaging, including plastic films which pack far more efficiently in a landfill than do many other waste components.

A recent example of this problem is the decision by McDonald's to abandon the foamed polystyrene clamshell for hamburgers in favor of a paperboard construction. The company obviously concluded that the efforts of polystyrene producers to establish recycling centers, recycling technology, and plants to recycle foamed polystyrene articles were inadequate to counteract the public's belief that these articles are landfill-unfriendly and dangerous to incinerate. While it is not evident that this widely publicised decision will negatively impact the use of plastic packaging films, it certainly will not improve the image of plastic packaging in people's minds and it will also encourage further action by those who would like to see plastic packaging curtailed.

The purpose of this discussion is not to advocate the selected banning of rigid plastic containers in order to preserve the film business—quite the contrary. Recycling and incineration of these packages along with many other solid waste components are feasible and energy effective ways of diverting them from overcrowded landfills and represent the most sensible solution to the landfill crisis. However, until alternative disposal practices such as these are widely adopted, and that time lies many years in the future, film manufacturers, converters and packagers must all face the fact that the unpopularity of plastics, including plastic films, among a small but highly vocal minority will force them to adopt defensive countermeasures.

Such countermeasures will include mounting recycling programs without regard to their economic merit; attempting to design films and film constructions which are easier to recycle even though such constructions may be less functional and more expensive than the constructions they replace; using additives in films which purport to enhance their degradability even though landfill excavation has shown that few materials commonly believed to be degradable do in fact degrade in a dry, well managed landfill; and, as a last resort, using either less plastic in packages or substituting paper when that is feasible, or both.

Not all plastic films will be equally affected by these developments. Those made from chlorinated polymers such as PVC and PVDC will be more strongly impacted because there is a widespread belief, which has been scientifically disproven but not dispelled, that incineration of these plastics results in the release of unsafe quantities of dioxins into the atmosphere. While there is no recorded evidence of fatalities caused by dioxins, they can cause chloracne and are certainly not desirable atmospheric components.

PVC in particular, first because it is a much more common packaging plastic than PVDC and second because of lingering fears about the harmful effects of vinyl chloride monomer, will probably suffer more even though there is no rational basis for that outcome. Polystyrene foam products are under intense pressure because of their low density and because some are still foamed with chlorofluorocarbons which adversely affect the Earth's ozone layer. Some replacement of these materials or other plastics is inevitable.

As this book is written, the major thrust of governmental action is to "do something" about rigid plastic containers. Films, other than trash bags which are attacked in some quarters because they do not degrade in landfills, have been relatively—but only relatively—free from attack. This respite is only temporary. Recognizing this, some manufacturers of trash bag films are using additives that supposedly enhance biodegradability. Corn starch manufacturers are developing polymers based on starch which they claim will be consumed by bacteria after being landfilled. Manufacturers of plastic ring connectors for beverage bottles and cans have for some time been using photodegradable films made from an ethylene-carbon monoxide copolymer which has mechanical properties that closely resemble LDPE. These films break into small fragments under the influence of wind and ultraviolet light. Nevertheless, at this writing, the biodegradability trend appears to be diminishing in favor of sounder approaches such as recycling, source reduction, and incineration.

This diminishing trend may accelerate as a result of a very recent study[1] that showed that in specially created, intensive bacterial environments, where paper virtually disappears in one day, most allegedly biodegradable plastics do not degrade. At best, these plastics only break down into smaller pieces as a result of bacterial attack on the starch or sugar components that have been incorporated in the polymer chain. The only plastic that is heavily attacked is a very expensive specialty material that is not yet cheap enough to enable use in environmentally sensitive products such as trash bags.

Conclusions

The U.S. will finally master this problem. The solutions adopted will have the following consequences:

(1) The growth rate of plastic packaging will slow—for a time.

(2) Products such as trash bags and disposable diapers will to some ex-

---

[1]Anon. 1991. "The New York Times," (March 5):C4.

tent be made from polymers that degrade slightly faster than those now used.

(3) The U.S. will incinerate 25–30% of its solid waste. Certain disposable products, such as batteries, will be largely excluded from the incinerator feed stream. All chlorine-containing polymers will lose a minor share of their market to chlorine-free polymers.

(4) Somewhere between 10 and 30% of all plastic packaging will be made from recycled polymers and used in non-food contact applications.

(5) Package constructions that lend themselves to recycling will be developed and finally represent perhaps 10% of the constructions now used. In this connection, a minor shift from multicomponent to single component constructions will take place.

SOCIOLOGICAL TRENDS THAT IMPACT PACKAGING

Major changes have taken place in U.S. society since World War II. Families have become smaller, older citizens, frequently single, are more often living alone, and women have become an important component of the work force. These changes, along with the development of the microwave oven, the much greater frequency of food consumption in restaurants, the rising popularity of ethnic foods, the growing health food trend and the overall consumer desire for greater convenience in food preparation, have led to major changes in the way that food and other retail articles are packaged.

Plastics and plastic films have been the largest beneficiaries of these changes and have grown much more rapidly than any other class of packaging materials with the possible exception of aluminum foil.

There seems to be no reason why these trends should shift *qualitatively* in the future, although some, such as the percentage of working women, must inevitably slow in growth *rate* as the supply becomes exhausted. Thus we expect the trend to more and more convenience packaging to continue. Technical developments in plastic films, discussed more fully below and driven in part by these sociological trends, will produce better and even more convenient packaging and continue to increase the dominance of plastics in these applications.

ECONOMIC TRENDS

The cost of packaging depends on four major factors: (1) the cost of energy, (2) the cost of raw materials, (3) the cost of labor and (4) the cost of investment—both equipment and working capital. In turn, the cost of materials depends in part on the cost of energy, as does the cost of equipment. Of these four factors, the trend in the cost of energy is the one for which a steady increase is the easiest to predict with any degree of certainty and one that will probably increase most rapidly compared to the others. Therefore this discussion of economic trends will focus on energy and particularly on how rising energy costs will affect the balance between flexible and rigid packaging and the relative positions of the three major flexible packaging materials: plastic films, aluminum foil and paper and wood pulp-based products such as cellophane and glassine.

Distribution Costs

The most obvious way in which rising energy costs affect the balance between rigid and flexible packaging is in distribution costs. The weight of metal or glass packages ranges from five to ten times the weight of flexible plastic or paper packages. This increased weight translates directly into increased consumption of vehicular fuels derived from crude oil, since both trains and trucks now run almost exclusively on diesel fuel. Of all the developed energy sources, crude oil is being depleted most rapidly, and although the price of oil has gyrated sharply up and then down in the past twenty years, it is inevitable that over the long term it will trend upwards as supplies slowly diminish. This will continue the pressure for weight reduction on all distributed products and the packages that many of them require. That pressure in turn will increasingly tip the scales in favor of lightweight flexible packaging.

Rising distribution costs will affect packaging materials and techniques in another more subtle way. For years, the trend in food production has been to concentrate farming, harvesting and preliminary packaging of food for bulk shipment in large facilities that are distant from major markets. From these facilities, food is shipped in bulk to processing locations closer to the ultimate consumer. In these locations, the bulk shipped product is repackaged, sometimes after further processing, into smaller retail packages that are much heavier per unit weight of contents than the large bulk packages used for long haul shipping.

This shift to dispersal and proliferation of final retail packaging operations benefits plastics and plastic films in two ways. First, bulk shipment, when not in tank cars or tank trucks, takes place more and more in large flexible plastic containers such as the bag-in-box type discussed in Chapter 6. Second, the final retail packager will often choose a flexible package if possible since the investment in flexible packaging equipment is far lower than the investment in new plants to make glass jars or metal cans.

The Cost of Materials

The total energy required to produce a pound of paper and a pound of plastic resin are about equal at 18,000 BTU/lb of final product. Conversion of resin to film requires another 1,000–2,000 BTU/lb. The energy required to produce a pound of aluminum foil from ore is far higher, in the range of 100,000 BTU/lb. These figures include the energy required to obtain and transport the raw materials: crude oil, natural gas, wood, and aluminum ore, as well as the energy required to convert those raw materials to their respective finished products.

Rising energy costs will continue to lead packagers to use substitutes for aluminum foil such as metallized films and will not affect the *relative* cost of plastic and paper-based products. It most certainly will affect the rigid/flexible packaging balance, since:

(1) The energy required to produce a pound of steel strip or aluminum ingot is two to four times greater than that for paper and plastic, and

(2) The lower density of plastics and paper, as compared to steel, aluminum, and glass, and the much lower packaging material consumption involved in flexible pouches vs. rigid containers make the energy cost *per package* ten to twenty times lower when flexible packaging is used.

The cost of materials involves more than the energy required to produce them. In the case of plastic vs. paper, it involves the price of crude oil or natural gas vs. the price of wood. Regardless of today's prices for these materials, it seems clear that the renewability of wood as compared to the non-renewability of oil and gas will lead to greater price escalation for the latter over the long term. Waste recycling will not have a major impact on this conclusion since it is likely that when recycling technology for plastics and paper are mature, the fractions recycled, now far higher for paper, will eventually be roughly the same for both materials.

In connection with relative raw material costs, there is one issue that has caused widespread confusion in the public mind. It is widely believed that since plastics are made from petrochemicals derived from crude oil or natural gas, the dwindling supply of these non-renewable resources will ultimately lead to the disappearance of plastics entirely from all but a few highly essential applications for these versatile materials. There are two reasons why this scenario is invalid:

(1) Plastics are the highest value-added large volume product made from oil or gas. Thus, as supplies of oil and gas gradually decline, fuels, which are lower value-added products, will be replaced by other fuels, first in stationary power plants that can use coal or nuclear en-

ergy, and ultimately in fuels for transportation, for which non-oil derived alternates such as methanol or ethanol are available.

(2) Plastics for packaging consume only about 1/2% of the oil and gas currently consumed in the U.S. Thus any program to avoid the use of plastics in packaging will have no discernable affect on the rate at which industrialized societies are forced to adopt alternative sources of fuel.

Other Economic Factors

The differences in labor-intensiveness between metals, glass, paper, and plastic are not large enough to be significant in any projections of relative future costs. Investment *per unit area of packaging material* in plants to produce these materials and packages therefrom is much lower for plastics and paper due in part to their much lower densities. Thus these two cost factors, insofar as they have any appreciable influence over cost trends, would appear to be most favorable for the lightweight materials: paper and plastic films.

Conclusion

Rising energy costs will continue to favor flexible over rigid packaging, plastic films and paper over foil, and may cause a minor shift in the paper/plastic balance in favor of paper. The latter effect cannot be large since paper has only a few of the many packaging-friendly attributes of plastics.

TRENDS AND POSSIBILITIES IN TECHNOLOGY

The two words in the title of this section—"trends" and "possibilities"—reflect the fact that some of the future changes in plastics technology can be predicted with some confidence because they have already begun and seem to be both viable technically and desirable from a cost and/or consumer point of view. These therefore can be treated as trends with a reasonably high degree of confidence that they will continue. Other changes are more difficult to evaluate from the technical feasibility point of view and so must be regarded as possibilities, meaning they *could* happen, given the right combination of inventive skill, reasonable cost and market acceptance.

Trends with High Success Probability

The development of new and modified polymers that will find applications as packaging films will continue. The major attributes of films based on these new polymers will be enhanced barrier to oxygen, greater stiffness and higher strength. The expanding research into liquid crystal polymers may lead to plastic films with some of these attributes, since these new polymers are already known to produce high strength and stiffness in thin sections. However, before films made from liquid crystal polymers can become viable in packaging, their cost must be reduced by at least a factor of ten. This will be a difficult task and the probability of success via this approach is therefore low.

The two high oxygen barrier polymers now available—EVOH and PVDC—each have a major drawback. PVDC is prone to decomposition at extrusion temperatures, and EVOH loses oxygen barrier rapidly as ambient humidity increases. Although we will see some improvements in these two polymers, it is unlikely that anyone will invent an entirely new barrier polymer that incorporates the best features of PVDC and EVOH without their drawbacks.

The continuing creation of more highly engineered films will lead to packaging films that combine all the properties needed to perform the multifarious packaging functions more efficiently and at lower cost. Certainly coextrusion will be an important factor in creating these lower cost, more highly engineered films, but developments in coatings and other film treatments will also play a role.

As two examples of the latter, we will eventually see:

(1) films with inorganic coatings, probably based on silica, which have gas barrier properties far better than PVDC and which will be more amenable to recycling than are composite films containing PVDC, which tends to decompose when recycled

(2) films with surface treatments that facilitate printing with solventless inks, thereby eliminating an air pollution problem that plagues many converters

(3) coextrusions that can be surface printed and still retain the appearance that is now possible only when the print is buried. This in turn will further accelerate the application of lower cost coextrusions in many food packaging applications where print quality is crucially important from a marketing point of view.

In the discussion of produce packaging in Chapter 6, the technique of controlled atmosphere packaging was described. The use of CAP techniques will grow and be greatly facilitated, even revolutionized, by the development of lower cost films with controlled permeability tailored to the wide variety of fresh and prepared products whose shelf life can be extended many fold by the use of this technique.

Low cost, high speed wrapping of citrus fruit in shrink and/or stretch films to prevent moisture loss will become a reality, allowing, for example, oranges and grapefruit grown in Israel to be shipped to Omaha and consumed as juicy fruit many months later.

The growing trend to single service type packaging of toiletries in the Far East, a situation where the customer buys only his or her daily needs, will spread rapidly outside the U.S., creating an ever larger market for plastic pouches, but is unlikely to penetrate the U.S., where the custom is to buy in large economy sizes. Packaging of the daily ration of cooking oil in India and Latin America was discussed in Chapter 6, where the technical hurdles and the large potential market were described. This trend, still embryonic, is certain to develop into a very large plastic pouch market in those regions but again, not in the U.S.

One of the largest consumer frustrations with plastic film packaging is the difficulty of readily penetrating the strong hermetically sealed packages that modern technology has made available to the packager. This will inevitably change, probably through the further development of modified resins and resin blends. In ten years, virtually all plastic pouches will be openable with only modest effort, since the market share prize for the innovative packager who is first out with a truly easy-open package will be too great to resist. Once that rather modest technical hurdle is surmounted, the more difficult problem of reclosability, preferably hermetic reclosability, still remains, and it is likely although not highly likely that hermetically reclosable pouches of this kind will appear with a ten year period.

In this connection it will be interesting to watch the evolution of blister and skin packages, both notoriously difficult to open. The inevitable advent of easy-open hermetically sealed pouches that will have food as their major market may not affect the blister and skin package since those are found mainly in non-food packaging. Thus the pressure for easy openability in non-food cases, while still great, may not intensify with the advent of the easy-open pouch. After all, various mechanical pouch closures of the zipper variety are available today to the non-food small item pouch packager but are infrequently used because of cost and because blister and skin packages offer the many merchandising attractions of carded display. Finally, reclosability is rarely an issue in non-food packages.

Future Possibilities

Developers of food packaging resins have long dreamed of finding the so-called "one-bag resin", a term used to describe the resin that at low cost can meet all the requirements for the many food items that must be packaged in gas barrier films. This is a major technical and marketing challenge and can therefore be regarded here as only a possibility. Films made from such a resin would have to be strong, clear, stiff, heat sealable and provide oxygen and moisture barrier comparable to that now furnished by PVDC, all at an area cost little higher than OPP. Perhaps the most compelling motivation for inventing such a resin is the enormous advantage it would have over multilayer films in ease of recycling.

The one-bag resin would probably not be highly popular with those converters who now add major value to plastic films by producing multilayer structures; their function would revert to simply printing film, an activity where competition is rife and there is far less scope for technological differentiation in the marketplace.

An important additional obstacle that the one-bag food packaging resin must surmount stems from the likelihood that such a product would be controlled for a lengthy period by patents held by one or at most a small group of resin companies. Such a situation is highly undesirable for both converters and end users, both of whom can readily imagine the predatory price policies that the monopolist or oligopolists will pursue once the packager has converted an entire product line to this new film. The fact that such price behavior on the part of the one-bag resin producer would be suicidal in the long term market is rarely regarded as a compelling argument by most large food processors.

In non-food packaging, as we have seen, the one-bag resin is commonplace: LDPE pouches, PVC carded blister packaging, ionomer skin packaging, stretch wrapping with EVA or LLDPE and many others are all "one-bag" situations.

The sterilization of food by the use of techniques other than heat has been an attractive but elusive goal for food processors striving to retain fresher natural tastes in their products. Radiation sterilization was regarded as inevitable thirty years ago but has made little progress since. Had radiation sterilization become popular, plastic packaging would be quite different today since gamma radiation will crosslink, discolor and embrittle many plastic packaging resins. Microwave sterilization is a newer option for food processors. One day it may become widespread, partially because the human safety concerns on which radiation sterilization foundered will surely not be an issue for a cooking process that already is used in more than 70% of American homes. Since only plastic and paper among flexible packaging materials are transparent to microwave radiation, these materials will obviously benefit from such a development. Nevertheless, the very large new investment required for food processors to convert from retorting to microwave sterilization will be a major deterrent to its widespread adoption, particularly since the taste advantages achieved are likely to be subtle and the upheaval in traditional packaging that will be required will not be an attractive prospect for many food companies.

Aseptic processing, in which the food and the container are sterilized separately, offers similar taste enhancement prospects and gives the processor additional container material options while taking away none. Yet here too the pace of the transition away from conventional retorting has not been rapid. Aseptic processing equipment that will handle foods other than simple liquids is now available and in some cases has led to the adoption of packaging methods that could not withstand retorting conditions, most notably the paper-based Brik-Pak described in Chapter 6. Plastics will continue to benefit from the further adoption of aseptic processing systems, but probably in the form of rigid plastic containers rather than as plastic films.

SUMMARY

In summary, it appears that, as noted in the introduction to this Chapter, the technological trends and possibilities that can be foreseen seem favorable to the continued growth of plastic packaging films.

Viewing the future more comprehensively, we are now seeing the collision of major forces acting in quite different ways to influence the future of plastic films: the powerful and basically anti-plastic environmental/ governmental/waste disposal movement that brings increasing pressure against growth, opposed by seemingly irresistible economic, sociological and technological pro-growth forces. The authors of this book believe that these positive growth forces will in the end prove too powerful to be contained.

APPENDIX

THE PACKAGING INDUSTRY

Introduction

This section describes how the plastic films and resins described in earlier chapters are brought together with the packaging machinery described in Chapter 5 and the various products discussed in Chapter 6 to create packaged products and distribute them to customers.

Five major industries are involved in creating a product in a flexible package (see Figure A.1):

- plastic resin and film producers
- flexible packaging converters
- packaging machinery manufacturers
- product manufacturers
- contract packagers

Also involved, but not discussed in detail here, are manufacturers of other flexible package components: cellophane, paper products, adhesives, aluminum foil, and printing inks.

Plastic Resin and Film Producers

All of the plastics used to make packaging films are manufactured from monomers made from crude oil or natural gas in large petrochemical refineries. Many of the large oil companies that refine these crude natural products and produce monomers are also the major producers of the plastic resins used to make packaging films. They have assumed this role

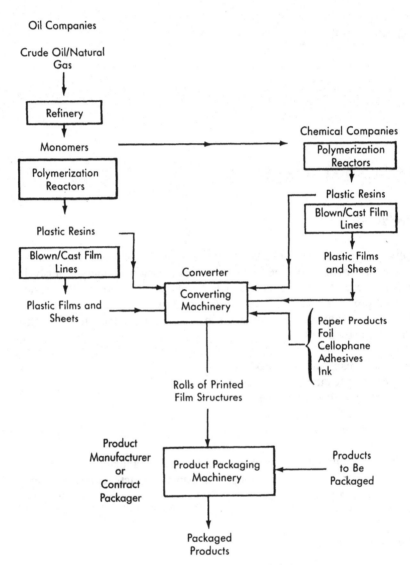

Figure A.1. Industry Structure and Material Flow.

because they found it advantageous to produce plastic resins in addition to their more traditional products: fuels for heating, power generation, and transportation. There are exceptions to this rule—the oil companies tend to dominate production of the very large volume commodity plastics such as LDPE and HDPE, while manufacture of the smaller volume specialty plastics such as PC or PET is more frequently the province of chemical companies who buy their raw materials from the oil companies.

Flexible Packaging Converters

The original function of the flexible packaging converter was printing the paper, glassine, and cellophane that were then the only flexible packaging materials available. As additional flexible packaging materials such as plastics and aluminum foil came on the market, converters expanded their function to include the creation of multilayer structures, mainly for food products. Early in this evolution, monolayer plastic films were produced solely by resin manufacturers who, unlike converters, had the technical resources to develop complex film manufacturing processes. Since converters had never made paper or cellophane, production of monolayer plastic films seemed to them at that time to be a logical function for resin producers rather than for converters.

As time passed however, some converters entered the plastic monofilm manufacturing business, making simpler films such as LDPE but still leaving the manufacture of more complex films such as PET, OPP, and BON to the resin producers. Today, monofilms that are relatively easy to make are produced by both converters and resin manufacturers, more difficult monofilms are produced by resin manufacturers, while film printing and multilayer film manufacture are the virtually exclusive province of the converter.

This arrangement sometimes puts the resin and film manufacturer in competition with converters who make monofilms, but resin manufacturers who make monofilms try to avoid such competition with their converter customers by concentrating their film marketing efforts on large volume, undifferentiated, unprinted films for products such as trash bags while converters who make monofilms concentrate their monofilm sales efforts on printed films customized for individual applications such as bread bags.

Contract Packagers

Rarely important in food packaging, contract packagers play an important role in non-food packaging, notably in the packaging of drugs and pharmaceuticals. These organizations perform solely a packaging function, serving product manufacturers who do not wish to invest in packaging machinery or who have relatively small amounts of product that requires a different type of package than they can make on the equipment that they have. This latter circumstance frequently arises when a manufacturer is introducing a new product and the package style and construction are not yet finalized.

Contract packaging businesses are fast, flexible, rapid turnaround type operations that typically can package products in a wide variety of ways. Most product manufacturers do not have the same equipment variety and usually are not nearly as flexible in their ability to convert from one package material/style to another at short notice.

Machinery Manufacturers

Three classes of machinery are required in the flexible packaging business:

- machines that convert resins to monolayer films
- machines that combine resins, films, adhesives, and ink to make printed multilayer structures
- machines that combine the product and the package to create the final packaged article

These machines are described in detail in Chapters 2, 3 and 5. The first two classes of machines are used by film producers and converters, while the third class is used by product packagers.

The flow of materials and products between these four industries is displayed in Figure A.1.

PROPERTIES OF PLASTIC FILMS

See Table A.1 on pages 233 and 234.

Table A.1. Properties of plastic films.

	LDPE	HDPE	LLDPE	>12% VA EVA	Ionomer	Oriented PET	OPP	PVC*	PS	PVDC	Nylon	EVOH**	BON
Density, g/cc	0.91–0.925	0.945–0.967	0.918–0.923	0.94	0.94–0.96	1.4	0.905	1.21–1.37	1.05	1.6–1.7	1.14	1.12–1.19	1.14
Yield, in²/lb-mil × 10⁻³	30	29.0	30.0	29.5	28.6–29.5	20–22	30.6	20–22.5	26	16.2–16.8	23.5–24.5	23–24	23.5–24.5
Tensile strength, kpsi	1.2–2.5	3.0 to 7.5	3.5 to 8.0	3–5	3.5–5.5	25	25–30	2–16	5.0–8.0	8–20	7–18	1.2–1.7	25–30
Elongation at break, %	225–600	10–500	400–800	300–500	300–600	70–100	60–100	5–500	2–3	40–100	250–500	220–280	70
Impact strength, kg-cm	7–11	1–3	8–13	11–15	6–11	25–30	5–15	12–20	N/A	10–15	4–6	N/A	N/A
Elmendorf tear strength, g/mil	100–400	15–300	80–800	50–100	15–150	13–80	4–6	N/A	N/A	10–20	20–50	N/A	N/A
WVTR, g-mil/100 in²-day @ 100°F and 90% R.H.	1.2	0.3 to 0.65	1.2	3.9	1.3–2.1	1.3	0.3–0.4	2.8	5.0	.05–0.3	24–26	High	12
Oxygen transmission rate, cm³-mil/100 in²-day-atm @ 77°F and 0% R.H.	250–840	30 to 250	250–840	515–645	226–484	5	110	5–1500	100–200	0.08–1.7	2.6	0.01	2

(continued)

233

Table A.1. (continued).

	LDPE	HDPE	LLDPE	>12% VA EVA	Ionomer	Oriented PET	OPP	PVC*	PS	PVDC	Nylon	EVOH**	BON
CO_2 permeability, cm³-mil/100 in²-day-atm @ 77°F and 0% R.H.	500–5000	250–645	500–5000	2260–2900	626–1150	N/A	240–285	50–13,500	N/A	0.04–10	4.7	N/A	N/A
Resistance to grease and oil	varies	good	good	varies	good	good	good	good	good	good	good	good	good
Dimensional change at high R.H., %	0	0	0	0	0	0	0	0	0	0	1.3	0	0
Haze, %	4–10	25–50	6–20	2–10	1–15	4	3	1–2	0–1	2	2	N/A	1–2
Light transmission, %	65	N/A	N/A	55–75	85	88	80	90	90	80–88	N/A	N/A	N/A
Heat seal temperature range, °F	250–350	275–310	250–350	150–300	225–300	275	200–300 (coated)	280–340	194–212	250–300	350–500	N/A	N/A
Service temperature range, °F	–70 to 180	–60 to 250	–60 to 180	–60 to 140	–150 to 150	–100 to 400	–60 to 250	–20 to 200	–80 to 175	0–275	–75 to 400	N/A	–100 to 400
Tensile modulus, 1% secant, kpsi	20–40	125	25	8–20	10–50	700	350	350–600	330–475	50–150	N/A	300–385	250–300

*Some PVC data depend on plasticizer content.
**Some EVOH data depend on ethylene content.

MARKET PRICES FOR PLASTIC RESINS AND FILMS

Table A.2. 1990 market prices for resins.

Resin	Price, $/lb
LDPE	0.40
LLDPE	0.40–0.45
EVA (<10%)	0.50–0.55
HDPE	0.45
HMW-HDPE	0.45
PP	0.35
PS	0.50
PVC	0.35
PET	0.60–0.70
Nylon	1.00–1.20
EMAA	1.10
Ionomer	1.25
PVDC	1.25–1.50
EVOH	2.15

Table A.3. 1990 market prices for plastic films.

Film	Thickness, mils	Price, $/lb	Price, cents/MSI
(Films are listed in order of ascending area price)			
LDPE	1	0.70	2.33
LLDPE	1	0.65–0.75	2.35
HDPE	1	1.00	3.5
OPP	1	1.00–1.10	3.5
PET	1/2	1.65–1.85	4.3
Ionomer	1	1.25–1.35	4.5
BON	1/2	2.75–2.90	5.9
OPP-coated	0.7–0.9	2.30–2.40	6.2
PVC	1	1.35–1.45	6.5
PS	1	1.75–1.85	6.9
PVDC-coated cellophane	0.7	1.95	7.5
NC-coated cellophane	1	1.75	9.6

The prices quoted in Table A.3 represent actual prices that would be paid by a large buyer, rather than list prices. They are reasonably accurate as of September, 1990, but will probably be out of date within 6–12 months. However, the relative values should be reasonably accurate for several years.

The film thicknesses chosen are either those that are the most widely used for each of the films (PET, BON, OPP cellophane) or are quoted at 1 mil to give the best comparison of cost per unit area. Buyers of film pay more attention to cost per unit area than they do to cost per pound, since it is actually area, or coverage, that they are buying. The figures in the last column represent the price in cents per thousand square inches. These figures give the best measure of the relative cost of these films to a large customer. The area price figures have been calculated from the unit price as follows:

LDPE has a density of 0.91, which gives 1 mil LDPE film an area coverage of about 30,000 in²/lb. Its price is 70 cents/lb. 70 cents divided by 30 gives 2.33 cents/MSI. The comparable area price for 1/2 mil PET, for example, can then be calculated from its density (1.4) and its unit price ($1.75) from the equation:

$$\text{PET area price} = \text{density ratio} \times \text{LDPE area price} \times \text{unit price ratio} \times \text{thickness ratio.}$$

Multiplying the LDPE area price (2.33 cents) by the density ratio (1.4/.91) adjusts the price to allow for the higher density (lower area coverage per pound) of PET. Multiplying by the unit price ratio (1.75/.7) adjusts the price for the higher unit price of PET. Then multiplying by the thickness ratio (1/2 mil PET/1 mil LDPE) reduces the PET price to allow for the greater area coverage of 1/2 mil PET as compared to 1 mil LDPE.

GLOSSARY

acid copolymer   A copolymer that incorporates, as one comonomer, an unsaturated organic acid.

abrasion resistance test   A variety of techniques are used to determine the degree of abrasion of a film surface using standard abrading materials under carefully controlled loading and temperature. Measures of abrasion include loss of weight and clarity.

acrylic(s)   A generic term for a family of polymers based on acrylic acid or derivatives thereof. When used in the packaging context, this term usually refers to these polymers being used as adhesives or coatings.

air knife   A device for providing a high-velocity curtain of air used to pin plastic films to quench rolls or to remove excess coating bath from an application roll.

amorphous   A plastic or a domain in a plastic consisting of polymer molecules packed in a random fashion. This term is used to distinguish from a crystalline plastic or a domain in which the polymer molecules are arranged in regular patterns.

anhydride graft   The product of a process (grafting) whereby a molecule containing an anhydride group is chemically attached to a polymer molecule. The anhydride group, when grafted to polyolefins, improves their adhesion to polar molecules.

anhydride group   When two adjacent COOH groups on a molecule lose a molecule of water between them, an oxygen-bonded linkage forms between the two groups to produce an anyhdride group.

annealing   The application of heat to a formed or oriented plastic article to relieve stresses resulting from the forming or orientation process.

anti-block   The name for a treatment applied to plastic film surfaces to keep them from sticking together ("blocking") when they are tightly rolled up on a mandrel.

aromatic   A hydrocarbon that contains benzene rings.

aseptic   Free from septic matter or disease-producing bacteria. In food processing and packaging, this is an adjective that describes the system used to package food in a sterile fashion.

atactic   Describes a polymer molecule with random placement of side groups along the polymer chain, as in atactic polypropylene.

barrier resin   Any one of a number of polymers that are relatively impermeable to oxygen, such as PVDC, EVOH, PVA, nylon, etc.

benzene   An aromatic hydrocarbon, C_6H_6, in which the carbon atoms are arranged in an unsaturated six-membered ring.

biaxial, biaxial orientation   A process for orienting a plastic film in both the transverse and the machine directions.

biodegradable   The tendency of a material to degrade under the influence of environmental factors such as moisture, oxygen, bacterial action, etc. Frequently improperly used as a catchall term to characterize all degradation mechanisms.

blowup ratio   In a tubular film process, the ratio of the final blown film diameter to the initial cast tube diameter.

BON   Abbreviation for biaxially oriented nylon.

butene   A straight chain hydrocarbon containing four carbon atoms and one double bond.

CO_2 permeability test   See OPV.

calendaring   A process in which a plastic mass is fed into the nip between two rolls that squeeze it into a film. Passage through subsequent sets of rolls produce a film whose thickness is defined by the gap between the last pair of rolls.

catalyst   A material that accelerates a chemical reaction but that is not consumed in the process.

cellophane   A clear film made from wood pulp.

center winding   A method of winding film rolls in which the core is driven.

chub   The name for the package used for some varieties of processed meat and certain explosive materials.

coating   A thin layer applied to a substrate (film, sheet, or container) either from a solution or emulsion of the layer material or by extrusion coating a thick melt layer on a cold substrate.

coating weight   The weight of a coating per unit area of the coated substrate after any solvent has evaporated.

coefficient of friction   A dimensionless number that expresses, for a given surface, the ratio of the force required to slide an object over a frictionless surface to the force required to slide the same object over the actual surface.

coex   Abbreviation for coextrusion.

coextrudable adhesive   Synonym for "tie layer" (see below). The adhesive used to bond together two otherwise incompatible polymers in the coextrusion process.

coextrusion   Simultaneous extrusion of more than one polymer layer.

COF   Abbreviation for coefficient of friction.

cohesive strength   Strength of the material *within* a film that resists forces tending to pull it apart, as contrasted with "adhesive strength", which is the resistance to forces pulling apart two films that are adhered together.

cold seal   Term used for an adhesive that can be used to create a seal without using heat.

comonomer   When more than one monomer is present in a polymer, all the monomers present are called "comonomers".

condensation polymerization   The formation of a polymer by the elimination of small molecules such as water, allowing the reacting molecules to join together in long chains.

converter   A manufacturer who uses raw materials such as plastic resins and films, paper, foil, cellophane, adhesives and inks to create rigid or flexible packages or packaging materials that are then sold to organizations who package products therein.

coordination catalyst   One of a particular class of polymerization catalysts.

copolyester   A polyester made from a mixture of two different glycols with an aromatic diacid or a mixture of two diacids with a glycol.

copolymer   A polymer formed from at least two different monomers. Strictly speaking, this term embraces polymers formed from any number of different monomers, but in common practice, the term usually applies to just two monomers.

corona treatment   Subjecting a polymer film to an electrical discharge to alter its surface characteristics.

covalent (bond)   The chemical bond in non-polar compounds that consists of a shared pair of electrons. Typical examples are carbon-carbon, carbon-hydrogen, carbon-chlorine, etc.

CPET   Abbreviation for "crystallizable polyester".

creep, creep resistance   When a body is subjected to a stress below that which produces rapid elongation, it slowly deforms, or "creeps". "Creep resistance" is the term that expresses the resistance of the body to this slow deformation process.

cross-linking   The joining of separate polymer chains by means of short links between them.

crystallite   A small crystalline area in a non-crystalline matrix.

cyclohexane   A hydrocarbon C_6H_{12} in which the carbon atoms are joined to each other in a six-membered ring with single bonds between each pair.

deadfold   When a film does not spring back after being folded, it is said to have "good deadfold". Most plastics have poor deadfold when compared to non-plastic films and webs.

degree of shrinkage   See dimensional stability.

diffraction   The tendency of radiation to produce geometric patterns when passed through an optical grating or a crystalline solid. The amount and pattern of radiation diffracted gives information about molecular size and structure.

dimensional stability (shrinkage) test   The percent change in film dimensions resulting from exposure to a standard high-temperature environment. The measurement is usually made separately for the MD and TD. For humidity-sensitive films, changes are also determined due to exposure to different levels of humidity.

dioxin   A complex organic compound. For its relevance to packaging plastics, see Chapter 7.

dispersion   As used here, a mixture of solid polymer particles in water.

double bond   When two carbon atoms in a molecule are joined together by two bonds instead of one, this term is used to denote those two bonds. See Chapter 1 for examples.

downgauge   Use a thinner film than had been previous practise.

draw (as in "melt draw")   To pull on a molten polymer, causing it to become thinner.

DSC   Abbreviation for differential scanning calorimetry.

DTA   Abbreviation for differential thermal analysis.

dwell time   The time a film being run on a packaging machine remains in contact with heat sealing bars.

EAA   Abbreviation for ethylene–acrylic acid copolymer.

elastic modulus   See "Modulus".

EMAA   Abbreviation for ethylene–methacrylic acid copolymer.

emulsion   A dispersion of one immiscible liquid in another or a dispersion of polymer particles in a liquid, with these dispersions being stabilized by a surfactant.

EP copolymer   Abbreviation for ethylene-propylene copolymer.

EPS   Abbreviation for expanded polystyrene.

ethylene   The molecule C_2H_4.

EVA   Abbreviation for ethylene-vinyl acetate copolymer.

EVOH   Abbreviation for ethylene-vinyl alcohol copolymer.

extrusion   The process in which a stream of molten polymer is produced by melting resin pellets and pumping the melt through a die by means of a screw rotating inside a heated barrel.

extrusion coating   The process by which a thin layer of plastic is applied to a substrate by squeezing the molten plastic through a die and onto a moving web of the substrate, where it thereupon freezes and remains tightly bonded to the substrate.

extrusion lamination   A process for joining two webs by feeding them through a machine that extrudes a thin layer of plastic between them to act as an adhesive. See Chapter 3.

FDA   Abbreviation for the U.S. Food and Drug Administration.

feedblock   A mechanical device that joins several polymers in a coextrusion process.

fibrillate   The tendency of a film, especially uniaxially oriented film, to split, when flexed, into narrow sections or fibrils along the direction of orientation.

flex crack   Cracking in a film produced by repeated flexing. See "flex life".

flex life   A temporal measurement of the endurance of a plastic film or article to repeated flexing.

flexographic printing   A flexographic printing press uses a flexible (hence "flexo") elastomeric material as the base to carry the design to be printed and to convey the ink from the ink reservoir to the surface to be printed.

flex test   (pinhole flex test) A tube of film is kept under gas pressure while it is rapidly flexed by moving end discs, to which the tube is attached, towards and then away from each other. The generation of a pinhole is detected by loss of gas pressure and the number of flex cycles to failure is measured.

(Gelbo flex test) The film sample is held at two opposite edges by clamps that move in opposite directions on parallel tracks, flexing the film each time they reverse direction. The number of cycles preceding catastrophic film failure is the flex life.

flexural modulus   See "Modulus".

fluidized bed   A mass of particles contained in a cylinder and kept agitated by a stream of gas fed in at the bottom.

form-fill-seal   See Chapter 5 for a description of this widely used packaging technique.

forming web   In thermoform-fill-seal packaging, the web that is heated and then formed into a cavity to hold the package contents.

fpm   Abbreviation for feet per minute.

freezer burn   The name for the dehydration that occurs when unpackaged or improperly packaged food is stored in the low-humidity atmosphere of a freezer.

ga   Abbreviation for gauge.

gas flush   The practice of flushing out ambient air from within a package so that more desirable gases can be inserted.

gas phase   Used in this book to denote polymerization processes that take place between gaseous, as opposed to liquid or solid reactants.

gauge   A term expressing the thickness of a film.

g/cc   Abbreviation for grams per cubic centimeter.

g-mil/100 in²-day   The unit that expresses how much water vapor, in grams, diffuses through a film of a given thickness (mils) per unit area (100 in²) per unit of time (1 day).

gel A general term denoting a separate polymer phase insoluble in a molten polymer or polymer solution and visible as a discontinuity in solid film. A gel can be a very high molecular weight or cross-linked version of the host polymer or a chemically different polymer.

glassine A derivative of paper.

gloss test Measured with an instrument that records the total amount of light reflected by the surface of a film with the incident light impinging at a standard angle less than 90 degrees.

glycol An organic molecule that contains two hydroxyl (OH) groups.

graphics The term used for the printed matter including illustrations that appear on packages.

gravure printing By contrast to flexographic printing (see above), in gravure printing a steel cylinder is used to carry the impression and the ink.

gusset A triangular section of a bag which facilitates the formation of a square bottom.

haze A measure of the "milkiness" of a film. Caused by light being scattered by surface imperfections or film inhomogenieties.

haze test Measures the amount of incident light that is not transmitted through a film sample.

HDPE Abbreviation for high density polyethylene.

head space The vacant area within a package above the level of the contents.

heat sealing Sealing a package by applying heat to cause the adhesive to melt.

heat seal range test Heat seal strengths are measured for seals formed at various temperatures. The heat seal range is defined by a lower temperature limit where the heat seal strength reaches an arbitrary minimum level and an upper limit where the film undergoes distortion or burn-through.

heat setting The process of annealing an oriented film to improve dimensional stability.

heat shrink, shrunk Application of heat to a heat shrinkable film will cause it to shrink and tightly conform to the shape of the object packaged.

hermetic Completely sealed by fusion so that no gas can escape or enter.

hexane A straight chain hydrocarbon containing six carbon atoms but no double bonds. Occasionally used in plastics technology as a foaming or blowing agent.

HFFS Abbreviation for "horizontal-form-fill-seal".

homopolymer The antonym of copolymer, i.e., a polymer made from a single monomer.

horizontal-form-fill-seal A form-fill-seal process operated horizontally as opposed to vertically.

hot tack The property of an adhesive or seal layer to resist forces that would pull the seal apart while it is still hot.

hot tack test   The force to pull apart a newly formed seal is measured by incorporating in the seals the open ends of C-bent pieces of spring steel of different stiffnesses.

hot wire sealing   Some plastic films can be sealed to themselves by the use of a hot wire instead of sealing jaws. A well made hot wire seal is much less obvious than, say, a fin seal and also uses less film.

hydrocarbon   A molecule composed solely of carbon and hydrogen.

immiscible   Used to describe two liquids that are not mutually soluble.

impact strength   The resistance of a film to breaking or tearing when subjected to sudden impact.

impact strength test   Measures the energy required for a projectile fired at high speed to penetrate a film sample.

index of refraction   A constant, characteristic of each substance, that is the ratio of the velocity of light as it passes through the substance to the velocity of light in a vacuum.

infrared spectroscopy   A process that measures the absorption by molecules of infrared radiation over a range of infrared wavelengths. Each chemically different molecule has a characteristic absorption spectrum.

institutional   Denotes a market for food that is other than the household market: restaurants, schools, hospitals, airlines, trains, boats, prisons, etc. In these situations, the food is prepared by institutional employees and served to the consumer, usually in groups.

intrinsic viscosity   The viscosity of a dilute solution of a polymer, which is frequently used as a measure of the molecular weight, or degree of polymerization, of the polymer. Often encountered as the abbreviation "IV".

ionomer   A copolymer with a constituent monomer that contains an acid group that is partially neutralized with an inorganic base, which replaces the proton on the acid group with a metal ion such as sodium or zinc.

isotactic   Describes a polymer molecule in which all of the pendant groups are arranged with the same symmetry along the polymer chain.

jaw release   A term that denotes the ability of a film to readily detach itself from hot sealing jaws once they have made the seal. Good jaw release is vital for packaging films used on high speed form-fill-seal equipment.

jaw release test   Measures the force required to pull a pair of heat sealed film strips away from the sealing jaws immediately after heat sealing.

lamellar   Describes a leaf-like molecular structure, usually found in oriented films where crystalline domains form in layers.

laminar flow   Flow of a molten polymer in which layers in the flow plane remain stable with no mixing between them. These layers can be imaginary or real layers consisting of different polymers or different viscosities or temperatures within the same molten polymer.

lamination   A multilayer film made by adhesively bonding two or more films together. Usually used to distinguish the structure from a coextrusion, which is a multilayer film made from several molten resins.

lap seal   Any seal made between two overlapping films. Used in contrast to "fin seal".

L/D   Abbreviation for length to diameter ratio.

LLDPE   Abbreviation for linear low density polyethylene.

LDPE   Abbreviation for low density polyethylene.

light transmission, %, test for   Percent light transmission is measured by measuring the light transmitted through a film from a source of known intensity.

machinable, machinability   A general term describing how well, or poorly, a film runs on a packaging machine.

mandrel   A cylinder, usually steel, around which film is wound.

MAP, MAP/CAP   Abbreviation for modified atmosphere packaging. MAP/CAP denotes the use of modified atmosphere packaging or Controlled Atmosphere Packaging, but never both, because the terms are mutually exclusive: "Modified atmosphere" means that the atmosphere in the package has been altered, but not permanently; exposure to the atmosphere will eventually restore ambient conditions. "Controlled atmosphere" means that the package is designed so that the modified atmosphere within is controlled throughout the shelf life of the package and not allowed to revert back to ambient.

master roll   The large roll of film wound during a film formation process which is normally slit into smaller rolls for later processing or shipment.

MD   Abbreviation for machine direction.

melt draw ratio   The ratio of the final thickness of a drawn melt to the initial thickness (usually the film die opening).

melt flow index   A measure of melt viscosity using an instrument similar to a capillary rheometer except that the piston is driven by weights.

melt strength   A term that expresses how much load can be sustained by a molten polymer. A high melt strength polymer can be drawn into a thin film at high rates as in extrusion coating.

melt viscosity test   Carried out with instruments such as a capillary rheometer.

metallizing   The process of applying an extremely thin metal coating to a non-metallic substrate.

methacrylic acid   An unsaturated organic acid used as a comonomer in the class of acid copolymers.

mil   One thousandth of an inch.

miscible   Two liquids are said to be miscible if they dissolve in one another in all proportions.

modulus   In packaging, used to denote the degree to which a film or sheet resists stretching before it reaches its elastic limit when an external stress is applied. A more precise but rarely used term would be "elastic modulus".

molecular weight   The weight of a molecule, consisting of the sum of the weights of its constituent atoms. Compared to most molecules, polymers have very high molecular weights.

monofilm   A single layer, homogeneous film.

monomer   A molecule which, when repeatedly combined with itself or other molecules ("comonomers") forms a polymer.

MVTR   Abbreviation for moisture vapor transmission rate.

MVTR test   Measures the weight loss from a cup of water with the test film sealed over the top.

neck-in   The amount of width loss that occurs as a flat molten film is drawn prior to quenching.

nitrocellulose   The product of the reaction of cellulose and nitric acid. Used as a coating for cellophane.

OPP   Abbreviation for oriented polypropylene.

optical density   A measure of the opacity of a metallized film layer. It is the log of the ratio of the intensity of transmitted light to incident light.

OPV   Abbreviation for oxygen transmission rate.

orientation   The process used to align polymer molecules in a film.

OTC   Abbreviation for "Over the Counter".

oxidation   Reaction of any substance with oxygen.

PAN   Abbreviation for polyacrylonitrile.

paperboard   A relatively stiff paper product, usually greater than 12 mils thick. This term is used to distinguish the product from paper, which is thin and quite flexible.

PE   Abbreviation for polyethylene. Used when the writer wants to avoid being specific about whether he or she is referring to low density polyethylene or high density polyethylene, but it is usually synonymous with low density polyethylene.

peel-seal   A package seal made using an adhesive that can readily be peeled open.

pentene   A five-carbon hydrocarbon containing one double bond.

permeability   A measure of the freedom with which gases or liquids can diffuse through a material.

PET   Abbreviation for polyethylene terephthalate. See "polyester" below and Chapters 1 and 2. In this book, PET refers specifically to an *oriented* film of polyethylene terephthalate.

photochemical process   A chemical process initiated by the absorption of light.

pigmented   A plastic that contains white or colored pigment incorporated in its structure.

plasma   A gas containing chemically active species such as free radicals or ions.

plastics   In common usage, articles or shapes made from polymers.

plasticized   A polymer, such as PVC, that has chemicals added to it to make it more flexible and more amenable to fabrication.

polar   Used to denote a molecule that has one end more positively charged than the other end. The antonym of "non-polar".

polyamide   Polymer made from a diacid and a diamine.

polyarylate   A polyamide whose acid and amine constituents are both aromatic.

polyethylene   Polymerized ethylene.

polyester   A polymer made from glycol and acid monomers by elimination of water. Used in packaging to denote polyethylene terephthalate (PET).

polymer   A high molecular weight molecule formed by reacting small molecules ("monomers") together to form a long chain consisting of many monomer units.

polymethylmethacrylate   Polymer made from methyl methacrylate monomer, which is the methyl ester of acrylic acid.

polyolefin   Any polymer whose monomer units are unsaturated straight chain hydrocarbons ("olefins") containing only carbon and hydrogen. Thus PVC and PVDC, which contain chlorine, PET, which contains oxygen and benzene rings, and PS, which contains benzene rings, are not polyolefins. The most common polyolefins are LDPE, LLDPE, PP, and HDPE.

polypropylene   Polymerized propylene.

polystyrene   Polymerized styrene.

polyvinyl chloride   Polymerized vinyl chloride.

polyvinylidene chloride   Polymerized vinylidene chloride.

PP   Abbreviation for unoriented polypropylene.

pph   Abbreviation for pounds per hour.

ppm   Abbreviation for parts per million.

primer, primed   The act of putting a thin coating on a substrate so that it will be more receptive to printing inks or adhesives.

PS   Abbreviation for polystyrene.

psi   Abbreviation for pounds per square inch.

puncture resistance test   Determines the force required to cause a standardized probe to penetrate a film.

PVA   Abbreviation for polyvinyl alcohol.

PVAc   Abbreviation for polyvinyl acetate.

PVC   Abbreviation for polyvinyl chloride.

PVDC   Abbreviation for polyvinylidene chloride.

quenching   Rapidly decreasing the temperature of a hot object, usually accomplished by plunging it in a bath of cold liquid or contacting it with a cold metal surface.

registration, registry   In printing, being "in registration" means achieving exact juxtaposition of images placed sequentially on top of each other to add different colors to the final composite image.

resin The term that denotes a polymer in the form of small pellets packaged in a bag or in bulk and shipped to the customer.

reverse printed The process in which a transparent film is printed backwards so that when it is flipped over, the printing appears right side up. When used in a package, reverse printed film will always have the printing ink on the inside where it is protected from scuffing and abrasion.

RH Relative humidity.

rheology The science of the flow and deformation of matter.

roll formation A general term denoting qualitatively how evenly, smoothly, and regularly film is wound on a roll.

rotogravure printing See "gravure printing".

run, runnable, runnability A film is said to run well on packaging machinery when it is easy to use and causes few machine interruptions.

seal strength test See "heat seal strength test".

seal temperature range test See "heat seal range test".

sedimentation The process of a particle or molecule moving through a liquid as a consequence of gravitational or centrifugal forces. The rate of sedimentation of a polymer molecule in a solution can be used to determine its molecular weight.

sheet flatness test Deviations from uniform flatness are observed by pulling out a length of film from a roll and using a standard tension to pull the film as flat as possible. Measurement of the amount of residual deviation such as in a drooped edge or gauge band is expressed in inches.

shrink temperature range The range between the lowest temperature at which 10% shrinkage occurs and an upper temperature at which the film distorts unsatisfactorily.

shrink tension test Measured as the force created when a film strip is held between two fixed clamps and heated to cause it to shrink.

skin package One that exactly conforms to the shape of the contents.

slip A measure of coefficient of friction (COF). High slip means low COF.

slip test The force required to cause one film surface to slide over another or a film surface to slide over a metal surface. Often measured using an inclined plane with a variable inclination angle.

slurry A suspension of solids in a liquid.

slurry polymerization Polymerization conducted between undissolved reactants.

spherulites Large spherically shaped crystal structures present in plastic film.

spunbonded, spunbonded polyethylene A film resembles woven cloth that is made by entangling and heat-welding filaments of plastic to form a strong, translucent web.

stearate   Any ester, or derivative, of stearic acid.

stretch film   Special grade of film manufactured for use on stretch-wrapping equipment. See Chapter 5.

subprimal   Term used in beef processing to denote cuts smaller than primal but larger than final.

substrate   A film to which subsequent layers or coatings are added.

surface printing   Printing on the outside surface of a package as opposed to one of the inside surfaces (see "reverse printing".)

surface winding   A method of winding film rolls in which the winding force is provided by the driven roll in contact with the surface of the winding roll.

surfactant   A surface active agent, usually a soap or detergent. When used in polymerization, it interacts with a suspended polymer particle and a liquid (usually water) to facilitate the formation of a stable dispersion of the particle in the liquid.

"Surlyn"   DuPont's trademark for its line of ionomer resins. As this book is written, "Surlyn" products are essentially the only commercially available ionomer resins in the U.S.

tackifier   A film additive that increases tack and cling.

tamper-evident   Any package that contains some device that reveals whether it has been surrepitiously opened and then reclosed.

TD   Abbreviation for transverse direction—the direction perpendicular to the machine direction.

tear strength test   (Elmendorf test) Determines the force required to propagate a tear already initiated by a cut on the edge.

tensile strength   The resistance of a plastic specimen to breaking when subject to a stress applied longitudinally.

tenter frame   Mechanical device used for biaxial orientation of film.

terpene   A class of hydrocarbons.

T_g Abbreviation for glass transition temperature.

thermal stripe, thermal strip   A stripe of adhesive applied to a film which allows sealing at the stripe location only, as constrasted to the more common practice of applying the adhesive uniformly across the entire width of the web.

thermoform   To form a sheet into a shape by using heat and pressure to force the softened plastic into a mold.

thermoform-fill-seal.   A process for continuously creating a plastic container by thermoforming the plastic in a die, inserting the food in the container, and sealing on a lid. Often abbreviated TFFS.

thermoplastic   A plastic that melts without decomposing. Most packaging plastics possess this characteristic.

thermoset Plastics that become hard when heated and will not thereafter melt without decomposing.

tie layer The common term for a coextrudable adhesive.

track, tracking A film that faithfully and without constant adjustment follows a desired path on a packaging machine is said to "track" well.

UHT Abbreviation for ultra high temperature.

ULDPE Abbreviation for ultra low density polyethylene.

uniaxial Used to denote orientation solely in the longitudinal direction, as opposed to biaxial (see above).

unsaturated Term for a molecule that contains at least two carbon atoms connected by two bonds rather than one.

UV Abbreviation for "ultra-violet".

UV catalyzed A process that is accelerated when exposed to UV light.

VA Abbreviation for vinyl acetate.

VFFS Abbreviation for vertical form-fill-seal.

vinyl Term for an organic molecule consisting of a carbon-carbon double bond with a polar group attached to one of the carbon atoms, as in vinyl chloride, vinyl acetate, or vinyl alcohol.

vinyl acetate An unsaturated monomer containing the acetate ester group.

vinyl chloride The molecule C_2H_3Cl.

viscoelastic A material that exhibits both elastic and viscous components of deformation and flow.

VLDPE Abbreviation for very low density polyethylene.

web A term that probably originated in the textile or paper industries, but that was adopted by plastic resin processors to denote a plastic film, not as a final discrete article, but usually as a long film somewhere in the processing stage, frequently being drawn off a large roll. For example, a "printed web" would be a large roll of printed film. "Apply the adhesive to the web" would mean "apply it to the film on a big roll".

weld To apply sufficient heat to melt a material in order to join it to another piece of itself or to another material. This is one common method used to create package seals.

weld line A visible line or scar formed by the merging of two polymer melt streams as in a film die.

wind-up The assembly of machinery used to wind up film on a roll after it is made.

WVTR Abbreviation for water vapor transmission rate.

INDEX

Printed in the United States
by Baker & Taylor Publisher Services